Wiesner, C. J.
Hydrometeorology

Hydrometeorology

Hydrometeorology

C.J. WIESNER

School of Civil Engineering,
University of New South Wales, Australia

CHAPMAN AND HALL LTD
11 NEW FETTER LANE, LONDON EC4

First published 1970

Set in cold type by E.W.C. Wilkins & Associates, London, and Printed in Great Britain by T. & A. Constable Ltd., Hopetoun Street, Edinburgh.
SBN 412 09870 9

Distributed in the U.S.A.
by Barnes & Noble Inc.

Contents

Preface

This book is written primarily for hydrologists and, in particular, those interested in the estimation of probable maximum precipitation. However there is much in it to interest others engaged in the environmental sciences as many chapters deal generally with our climate or environment and discuss weather phenomena, their causes, measurement and variations in time and space.

The treatment has been deliberately kept simple. Thus, most of the material is within the scope of the average undergraduate. A more complicated physical and mathematical approach is available through the references for those interested. As design and maximum floods are not usually treated until the later stages of a postgraduate course in hydrology, the latter part of the book on probable maximum precipitation is of main interest to postgraduate students.

It has always been a puzzle to me that engineers and scientists, who are concerned with studying and changing the natural environments the better to serve the needs of the community, rarely receive formal lectures on the nature and measurement of this environment, that is, on meteorology and climatology. Their efficiency as investigators and research workers might be improved by such courses. The absence of them is overcome in various ways, by using consultants, taking postgraduate lectures at certain universities, and by reference to books. However, there are few texts in this field of applied climatology or meteorology orientated towards the practising engineers. It is hoped this book will fill a gap.

When the question of a book on Hydrometeorology first arose, the term referred to the study of the atmospheric portion of the hydrological cycle and it is still used here with this meaning. For administrative and other reasons a wider definition was adopted by W.M.O. in 1963 but most technical workers still prefer to use the term in its first sense and see Hydrology as the more embracing term for the land phases of the cycle. An alternative title to Hydrometeorology for the book might well be 'Estimating the Probable Maximum Precipitation', but as two thirds of the book is of wider interest this could be misleading.

Although I agree with the statement made by Mr Vance Myers in UNESCO Water Resources Series No. 30, that 'the practice of Hydrometeorology has not been reduced to a handbook', there are broad principles which can be laid down. If we are to modify our environment and use estimates from meteorological data, we must understand this environment in a quantitative sense. I hope that this book will serve as a guide to methods although practical hydrometeorologists may find it vague. However, each meteorological situation has its individual solution just as each major engineering project is unique. In the same way sound judgement needs to be exercised and based on the principles laid down. The book aims more to give an understanding of the uses of meteorology in engineering than to train hydrometeorologists. It needs to be supplemented by studies in specific valleys and by practical experience.

The use of the SI and/or fps systems was considered early in discussions on the book and although the former is now accepted, it was decided to use both for the following reasons. The book will probably be used with data which are in the fps system in their original state, in official publications and on cards. Even though new records are in the SI system and conversion is proceeding rapidly, it is likely that the old records will be used for a long time to come and students and professionals will need to be conversant with both systems. In any case, conversion from one system to the other is simple. Furthermore, a mixture of both systems gives familiarity to those who have worked only in fps before and enables those working in SI to appreciate the data in the older form.

For engineering students, I felt it was undesirable to cover the complexities of meteorology and climatology. Those are dealt with in specialised texts. Consequently only the simpler aspects of the subject and those which are useful and of application to the problems in hand have been described. Perhaps this is incomplete and may offend the purists, but I see no point in burdening students with complex mathematics and physics which they might never use again. However, it is important that the student see the fundamental scientific reasons behind the natural processes and attempt to evaluate them. The numerous empirical relationships and techniques, the nomograms, the charts and weather maps, so dear to engineers and meteorologists, are good and useful tools as long as their basis and derivation is understood. Only then can they be used intelligently and improved. I support Merrill Bernard in his conviction (*Am. Soc. Civ. Eng.* Jan 1943) 'that an engineer is not prepared to deal adequately with precipitation, the principal hydrologic factor, until he has gained a working knowledge of the processes which take part from the sea, transport and precipitate over the land the myriad of rain drops constituting a great storm. Thus equipped he will find his concepts broadened, his techniques made flexible, and his judgement strengthened'.

I should like to acknowledge the great help given to me, in the preparation of this book, by my friends, colleagues and students. For fear of omitting some, I shall not list them. However, mention must be made of Professor P.O. Wolf, who first suggested the text, and my wife and children, who bore the brunt of the detailed work. I am also particularly grateful to the following organisations for permission to use material from their publications:

Environmental Science Service Administration.
United States Weather Bureau.
The Royal Meteorological Society.
The American Water Works Association.
American Society of Civil Engineers.
International Association for Scientific Hydrology.

C.J. WIESNER

CHAPTER ONE

The atmospheric phase of the hydrological cycle

1.1 Introduction. The hydrological cycle deals with the movement of water, in its three phases of gas, liquid and solid, from the ocean, land or living matter into the atmosphere by evaporation and transpiration. It passes through complicated atmospheric phenomena, generalised as the precipitation process, back to the earth's surface, upon and within which it moves in a variety of ways and is incorporated into nearly all compounds and organisms. This cycle is demonstrated in Fig. 1.1 where the main fields of study are also shown.

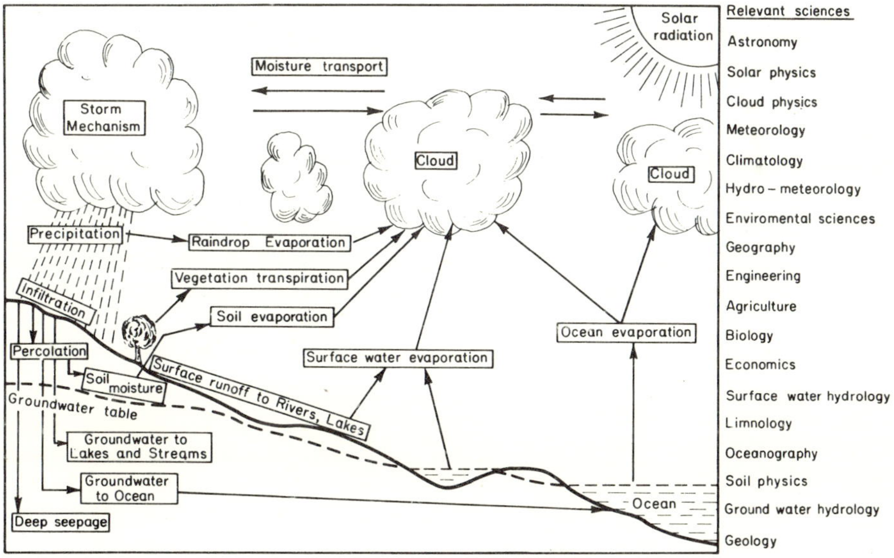

Fig. 1.1. The hydrological cycle and relevant sciences.

Every science, at some stage, is involved in part of the cycle and, although Hydrology may be defined as the science dealing with the hydrological cycle, it is necessary to define its limits and the various subjects with which it is allied.

1.2 The scope of meteorology. Meteorology deals with the atmospheric portion of the cycle and is often defined as the science of atmospheric phenomena. Meteorology is an observational science and a good set of observations is necessary for valid predictions.

Weather is more precisely described as the sum total of all the meteorological variables at a given time and a given place. These are measured qualitatively when types of clouds, storms or precipitation models are described, but more often they are measured quantitatively as with solar radiation, temperature, pressure, wind, moisture content of the air, and precipitation depth.

The continuous nature of the weather process and its many variations and relationships in time and space introduce difficulties of description, analysis and extrapolation. Although progress toward a better scientific description of weather is being made, it cannot yet be fully expressed in physical terms, nor are its many processes properly understood. In spite of its meagre data and incomplete knowledge, meteorologists, like engineers, are required to submit answers to problems. In most cases estimates of conditions are given rather than calculations and this should always be remembered by the users and originators of meteorological information.

1.3 Specialised branches of Meteorology. The science of meteorology touches on most of the activities of mankind and some of its applied fields have been given special names. *Meteorology* deals with the physical processes within the atmosphere or weather, as defined above. *Climatology* is the science of climate, by which is meant the collective state of the atmosphere at a given place within a specified period of time. *Micro-climatology* and *Micro-meteorology* consider the detailed variations of the meteorological elements and the physical processes near objects and surfaces. 'Climate near the ground' is a common and descriptive term of its principal interest. By contrast, *Macro-climatology* assesses the meteorological elements of the freely moving air, clear of local influences. *Industrial* and *Agricultural Climatology* or *Meteorology* deal with the data or weather processes which concern secondary industry and agriculture. For example, problems such as the siting of buildings, atmospheric pollution, and air conditioning concern industrial climatologists. Frost incidence, water requirements of crops, and the effects of weather on plant and animal disease all interest the agricultural meteorologists. *Hydrometeorology* can be defined as the study of the atmospheric processes which affect the water resources of the earth and, are of interest to the hydrological engineer. A broader definition of Hydrometeorology was given by the World Meteorological Organisation Fourth Congress (1963) [1] in stating that 'Hydrometeorology is concerned with the study of the atmospheric and land phases of the hydrological cycle with emphasis on the inter-relationships involved'. This embraces not only the atmospheric stages of the hydrological cycle but part of what is normally ascribed to surface water hydrology. The more restricted definition will be used in this text.

Applied Climatology or *Applied Meteorology* is a general term which covers the application of these sciences to human activities. Thus Industrial, Agricultural and Hydrometeorology are all facets of Applied Meteorology. There are many applications of meteorology, particularly in the engineering profession. If a project is considered under the stages of investigation and design, construction, and operation, [2] meteorology can make contributions in each phase. The hydrometeorologist, knowing the requirements of the engineer, can provide the correct type of data and present it in the best form.

1.4 Uses of Meteorology in the investigation and design stage. In the investigation and design stage of a hydrological project the meteorologist can:

(*a*) Provide direct climatological data. For example, he may give wind data for loadings on structures; temperature, radiation and humidity data for durability of materials; information on the type and intensity of storms for waterway calculations.

(*b*) Analyse data in sophisticated ways to the requirements of the project. For example, he may give information on the nature of the statistical distribution of the elements so that the frequency of occurrence of certain events can be estimated as well as the probable extremes. Extensive use of punch cards, tape and computers enables analysis to be tailored to the project.

(*c*) Estimate probable conditions when data is not to hand. Knowledge of the physical relationships between elements with time and space enables interpolation to be done. Isopleths of climatological elements are a simple means of interpolation. More complicated relationships can be established by statistical correlations and graphical methods.

1.5 Uses of Meteorology in the construction phase. The meteorologist can assist during the construction of a project by providing:

(*a*) Knowledge of average weather conditions and extremes, to plan a programme. Unfortunately, seasonal forecasts are unreliable but, where available, may be used with reservations.

(*b*) Short term forecasts and weather reports to enable the planning of day to day operations. For example, labour may be allocated to sheltered work if rain is forecast. The placing of concrete can be arranged to avoid frost or high temperatures. The compaction of soil may proceed when moisture levels are suitable or necessary modifications made.

(*c*) Special warnings at certain stages of the project. Vulnerable plants, material and labour can be protected if flood and gale warnings are received and acted upon.

1.6 Uses of Meteorology in the operational phase. Once the project has been completed, its efficient operation may be enhanced by meteorological advices. An obvious example is in the operation of multipurpose reservoirs where quantitative forecasts of rainfall depths enable appropriate releases or diversions from storages. Peak loads in an electrical supply system are closely related to weather and may be anticipated from accurate weather forecasts. Advance decisions sometimes have to be made on the basis of future weather. Long range forecasts give a small advantage of probability in such cases. This is better than no information at all, provided the limitations are realised and some assessment of probabilities is given.

1.7 Meteorological aspects of the hydrological cycle. In the atmospheric phase of the hydrological cycle the following factors are important to the hydrometeorologist and are discussed further in this book.

(*a*) The energy which maintains the cycle, that is, solar radiation and its distribution between the earth's surface, atmosphere and return to space as long wave radiation.

(*b*) The atmospheric circulation which develops from the receipt of solar radiation on a rotating earth within an envelope of air. It is important in distributing the heat energy over the globe.

(*c*) The effect of solar radiation upon the atmosphere and earth's surface to give the observed variations in the properties of the air, ground and water surfaces. The changes of temperature, water vapour content, wind and pressure with time and space are relevant.

(*d*) The evaporation process maintains the supply of moisture within the envelope of air and distributes the heat energy throughout the globe. Evaporation is a consequence of the receipt of energy and the change of state of water and can be estimated.

(*e*) The precipitation process is the converse of (*d*), and is concerned with the conversion of the water vapour back to the liquid or solid state. From the physical processes concerned, it is possible to estimate rates of rainfall in given situations and so the extreme rainfall intensity or the probable maximum precipitation.

Both the evaporation and precipitation processes effect the transfer of energy over the earth's surface and throughout the atmosphere. Energy is stored as latent heat in the evaporation process to be released on condensation and made manifest in the storm system.

1.8 Evaporation and precipitation over the globe. The occurrence of evaporation and precipitation over the globe depends upon the energy received and released. For example, if the annual amounts are considered, then evaporation is least in high latitudes and greatest between the tropics, although slightly lower near the equator where high humidities and cloud deplete the solar beam.

Precipitation is a maximum in the equatorial region due to the convergent air inflow from both hemispheres and the adequate supply of moisture. It decreases towards latitude 30° and then increases to a smaller maximum about latitude 60°. This precipitation pattern results from the general circulation set up consequent on the receipt of solar energy on the rotating globe and the need to distribute the energy over the globe and to maintain a balance. About latitude 60° air masses of contrasting properties interact with frequent precipitation. Energy, as latent heat in the warm moist air masses from the tropics is transported to these higher latitudes where it is released in the precipitation process.

The ocean is the source of most of the moisture which is precipitated over the land. It has been estimated that of the precipitation falling on the earth's surface, 35% of the moisture is derived from evaporation or transpiration from land areas and 65% from ocean areas. Of this precipitation, 25% is runoff and 75% infiltrates into the soil, from which it is evaporated or transpired into the air. Of this 75%, the wind circulation transfers 40% into the air over the ocean leaving 35% for land precipitation together with the 65% which moves in from the ocean. Figure 1.2 shows the movement of water vapour in this part of the hydrological cycle and the overall mass balance. Precipitation over the sea has not been considered nor the evaporation necessary for it.

1.9 Design in Meteorology and Hydrology. Engineering standards and recommended practices are specified on the basis of carefully collated and analysed data. Where experiments can be performed, these standards and practices are well documented and established. For example, various materials are subjected to laboratory tests and their strengths determined. Structures are loaded experimentally, and in hydraulics, the behaviour of water is observed in models of spillways, canals, dams, or land surfaces. Design in hydrology, involves such things as the specification of the water yield, flood potential, infiltration characteristics and groundwater flows of catchments. These are as necessary to the hydrologist as the strength of materials and the loads on structures are to the structural engineer.

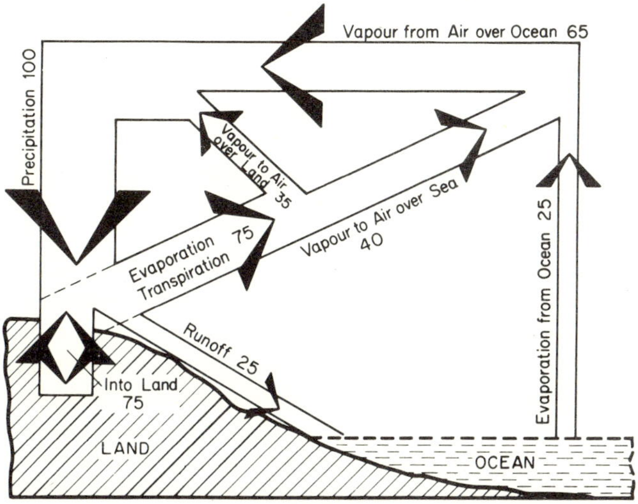

Fig. 1.2. Precipitation and evaporation balance.

In obtaining the necessary data for Hydrology and Meteorology, only a few of the natural processes can be simulated in the laboratory, in a simplified form, and mainly of academic interest. For example, the experiments of cloud physics examine the coalescence of cloud droplets. Usually, nature is relied upon to produce the necessary information and cannot be hurried. The average conditions in an area can be assessed in comparatively few years, the length of record required depending on the variability and nature of the element considered. However, estimations of the extreme values demand much longer records, in fact, as long as possible, and frequently, the required event may not have occurred and must be awaited. Alternatively, use can be made of the techniques of correlation, transposition and synthesis to extrapolate records.

1.10 Data collection and networks. The strongest emphasis is placed on data collection in hydrology and its allied subjects. Networks should be set up as soon as possible and be as complete and thorough as finances allow [3]. Such data collection is usually a fractional percentage of the works which may be initiated and can save very many times their own cost. First priority should be given to a permanent basic network which is comprehensively planned and equipped and collects data continuously and indefinitely. This network may be supplemented by denser networks, which are set up for shorter periods to obtain finer detail within an area where a project may be imminent. The data from these stations are correlated with the base stations and can be tied into the long term averages and trends. There are also special stations used for specific purposes such as research, power house operation or soil conservation work.

When extreme events occur, for example, record floods, intense rainfalls, high dew points and temperatures, every effort should be made to obtain all available information during or shortly after the event. The collection of private reports, the measurement of rainfall from containers left in the open, the marking of peak flood heights by debris, the observation of erosive damage, and information from local newspapers, radio stations, and inhabitants can materially assist in building up patterns of the magnitude and intensity of phenomena which may not occur again within a lifetime.

The optimum network is one which gives all the quantitative data, averages and extremes necessary to define the statistical distribution of the basic hydrometeorological elements or enable them to be interpolated at any point in the country.

The minimum network is more often the first step and is governed by available finance and resources. The principal stations are the first to be set up and good quality records must be obtained.

In any reorganisation, the existing stations should be retained, at least until satisfactory correlations are established. The older records should be tied into the reorganised network and an assessment made of their reliability.

In reaching decisions on the location of stations within the network and their density, attention must be paid to the nature of the elements measured, the purposes for which the data is required, the nature of the terrain, altitude, location, nearness to sea, climatic region, and population density. Although adequate sampling is the best method of determining the necessary network, this is rarely possible, and the network densities laid down by the World Meteorological Organisation might be best consulted [3]. For example, the minimum precipitation gauge network is given in Table 10.1.

1.11 Length of records required. The length of record necessary to obtain an adequate description of a meteorological element and to cover its variations in time, depends upon the same factors as the network density, with some differences in emphasis. Table 1.1 shows the number of years estimated to be needed [4].

Table 1.1. *Number of years of records required of meteorological elements taken at various locations to give a stable frequency distribution.*

Climatic element	*Extra tropical regions*				*Tropical regions*			
	Islands	*Coast*	*Plains*	*Mountains*	*Islands*	*Coast*	*Plains*	*Mountains*
Temperature	10	15	15	25	5	8	10	15
Humidity	3	6	5	10	1	2	3	6
Precipitation	25	30	40	50	30	40	40	50
Cloudiness or sunshine	4	4	8	12	2	3	4	6

REFERENCES

[1] World Meteorological Organisation (1963). *Fourth Congress* W.M.O., Geneva, Switzerland.

[2] WIESNER C.J., (1964). Hydrometeorology and river flood estimation. *Proc. Instn. civ. Engrs* **27**, 153–167.

[3] World Meteorological Organisation (1965). *Guide to Hydrometeorological Practices.* W.M.O., Geneva, Switzerland.

[4] Army Air Forces Headquarters (1943). Studies of length of record needed to obtain satisfactory climatic summaries for various meteorological elements. *Weather information branch Rep. No. 588,* Washington, D.C.

OTHER REFERENCES

ACKERMANN W.C., COLMAN E.A. and OGROSKY H.O., (1955). From ocean to sky to land to ocean. *"Water" The Yearbook of Agriculture, 1955,* U.S. Govt. Printing Office.

BERNARD M., (1944). The primary role of meteorology in flood flow estimating. *Trans A.S.C.E.* **109**, 311–382.

BROOKS C.E.P., (1951). *Climate in everyday life,* Philosophical Library N.Y.

GEIGER R., (1965). *The Climate near the Ground,* Harvard University Press, Cambridge, Mass.

LANDSBERG H.E. and JACOBS W.C., Applied Meteorology (1951). *Compendium of Meteorology,* American Meteorological Society, p. 979.

LINSLEY R.K., (1960). Common ground of Meteorology and Hydrology. *Bull. A.M.S.* **41**, 423–428.

CHAPTER TWO

The atmosphere and its heat balance

2.1 The atmosphere. The atmosphere is the laboratory of the meteorologist who studies its properties and the phenomena which occur within it and interprets them in terms of physical processes.

The atmosphere operates essentially as a thermo-dynamic engine in which the fuel is the energy from the sun. The gas laws hold within this system and enable many deductions to be made.

The permanent constituents of the atmosphere, by volume, are nitrogen 78%, oxygen 21%, and inert gases. The variable constituents may be in a solid, liquid, or gaseous form. Water is the most important of these, and can comprise as much as 4% of the atmosphere and exist in all three states. Ozone and carbon dioxide exist as gases in widely varying amounts. Solid components, besides water as ice, snow etc, are particles of dust, smoke, salts, and micro organisms. They reflect, scatter, and dissociate radiation and in particular light, to produce the various colours of the sky and sunsets. They also play a part in the condensation of water vapour, the formation of clouds and in the precipitation process.

The atmosphere has no clear-cut upper limit, but gradually decreases in density until it becomes indistinguishable at about 600 km from interstellar gas. There is sufficient air at 300 km to obstruct meteors and cause them to glow white hot. However, nine tenths of the atmosphere is below 20 km, and two thirds of it below Mt. Everest, height 9 km.

Vertical air soundings show that temperatures vary with height and this has given a division of the atmosphere into layers.

The troposphere is the layer within which the temperature normally decreases up to the *tropopause* where there is a change in the temperature lapse rate. The movement of water as liquid, solid or gas, is restricted to the troposphere where weather systems and phenomena occur. Water vapour moving in the troposphere is acted upon by weather processes to give precipitation on to the earth. The *tropopause* varies in height with latitude and season, ranging from 16 km in the equatorial region to 8 km over the higher latitudes. Tropopause temperatures vary from about 190°K to 230°K and are least in value in the tropics, where the tropopause is highest. Polar regions do not generally locate the lowest tropopause heights and warmest temperatures. Often these extremes occur over continents which extend into the artic zone. In the northern hemisphere these are found over Siberia and Canada. *The stratosphere* is the layer above the tropopause and is more uniform in temperature and in air movement.

Although weather is considered as resulting from the interchange of heat energy within the troposphere, processes above this layer giving ozone variations, ionisation, radiation filtering, and large scale air movements or waves, may influence weather conditions in the troposphere.

Near the earth's surface, the various meteorological elements vary with height and the zone where this occurs becomes a boundary layer with a transition of properties across it. These boundary or surface layers occur because of friction and thermal influences which show a daily or seasonal cycle. The depth of the layer may be specified by the size of the eddies which distribute the particular meteorological property throughout the layer. For example, in the lower 3–500 m, the vertical wind distribution is affected by large eddies generated by friction in the passage of air over the earth's surface and by thermal effects. Above 3–500 m, the wind flow is relatively free of these influences. The boundary layer for the wind field is thus below 3–500 m. Considering air temperatures in the lower part of the wind's boundary layer, it is observed that at night and more frequently in the colder seasons, the air temperatures sometimes increase with height or form temperature inversions instead of the more usual lapse rate of decreasing temperature with height. However during the daytime the wind and thermal eddies mix the layer to give consistent lapse rates and homegeneous temperatures and humidities. Thus representative temperatures and humidities are measured at about 1 m and surface winds about 10 m. Although these heights may not fully represent the meteorological properties of a wide area throughout the 24 hrs, it, at least, indicates the conditions where man lives and is a basis for subsequent interpolations.

Below 1 m, and in close proximity to objects or animals is the micro-meteorological zone throughout which a variety of frictional and thermal effects create small eddies and cause their damping. There is consequently a very wide range of temperatures, humidities, wind velocities and gas concentrations. The many changes of these elements with time and space are difficult to measure and relate to the macro-climatological conditions at 1 m. Most biological life which depends upon the earth's surface, lives and developes within the diverse conditions of this micro-climatological zone, and seeks or creates those micro-climates which suit them best.

2.2 Solar radiation and the planet Earth. It will be accepted that weather results from the interaction of solar radiation on the earth's atmosphere and surface. Therefore it is necessary to consider the earth's position and motions with respect to the sun, the sun's output, and the way this energy is disposed when it reaches the earth. The rotation of the earth on its axis gives the diurnal variations of the meteorological elements. There are the well-known variations over 24 hrs in temperature, humidity, pressure, radiation and local winds. Also, the earth revolves around the sun in an elliptic orbit with the sun at one focus and its equatorial plane inclined to the plane of its orbit at an angle of $23\frac{1}{2}°$. The two movements explain the changing elevation of the sun as well as the latitudinal and seasonal variations in length of day, receipt, and escape of radiation and weather.

Figure 2.1 exaggerates the eccentricity of the elliptic orbit and shows the equinoxes on March 21 and September 23 with the sun directly overhead at noon on the equator. The solstices are also shown as occurring on June 22 and December 22 when the sun is vertically above latitudes $23\frac{1}{2}°$, or the tropics of Cancer and Capricorn. At the time of the solstices the area south and north of latitudes $66\frac{1}{2}°$, or the Antarctic and Arctic circles are in darkness.

The earth is nearest the sun or perihelion at the time of the northern winter or southern summer. At this time approximately seven per cent more radiation is received at the top of the atmosphere near the equator. That is, on June 22, approximately 807 cal/cm^2/day are received compared with 860 on December 22. The seasonal effect is far greater than this and can be seen from Table 2.1.

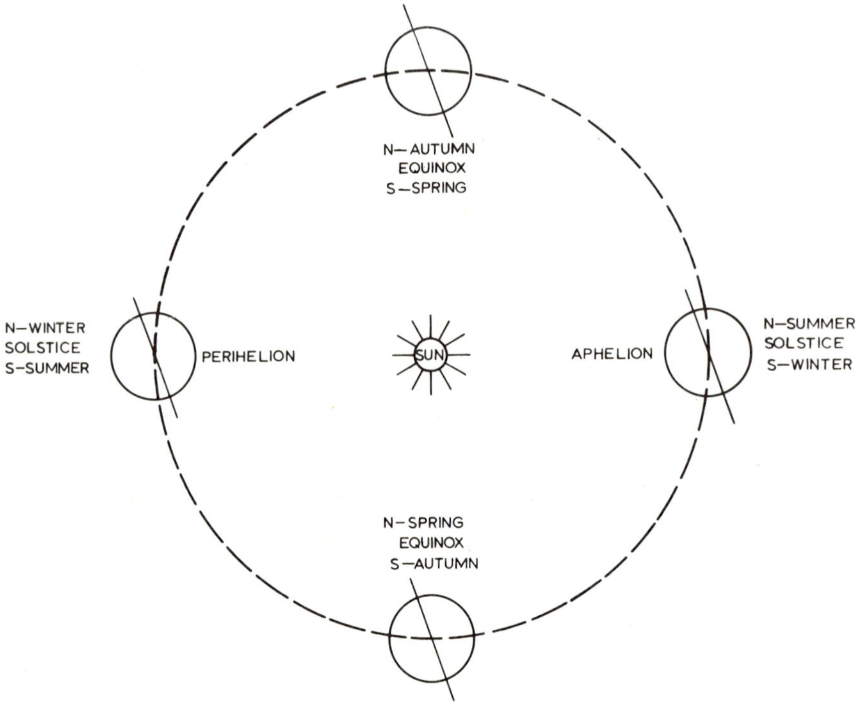

Fig. 2.1. The positions of the earth in its orbit.

Table 2.1. *Radiation at upper limit of the atmosphere in* cal/cm^2/day

Latitude	*Perihelion (Dec. 22)*	*Aphelion (Jun. 22)*
40°S	1080	303
40°N	323	1011
Equator	860	807

The sun at a temperature of about 6000 K° is the source of nearly all of our energy. The earth intercepts an infinitesimally small part of the sun's output, 5×10^{-10} %. Only a portion of the sun's radiation reaches the earth's surface as direct radiation, the remainder being reflected, absorbed, or scattered by the atmosphere. Considerable interchange of energy takes place between the sun, the earth's atmosphere, and the earth's surface and is discussed later.

The radiation transmitted from the sun is in the form of short electromagnetic waves which are screened by their passage through the ionosphere. By the time it reaches the stratosphere, 99% of the energy is contained between the wavelengths $0{\cdot}17\mu$ to 4μ with a maximum at $0{\cdot}475\mu$. Note that $\mu = 10^{-4}$ cm, some familiar wavelengths are violet light $0{\cdot}4\mu$, red light $0{\cdot}7\mu$, ultra-violet light less than $0{\cdot}4\mu$, and infra-red greater than $0{\cdot}7\mu$.

The atmosphere can selectively absorb electromagnetic waves with the amount of absorption dependent mainly on the length of path and the constitution of the atmosphere. Water vapour and carbon dioxide are particularly significant in this absorption. On the average, water vapour absorbs 11% and other gases 4% of the direct radiation.

The solar radiation may be measured by radiometers, actinometers and pyro-heliometers. These operate on the principle of the thermocouple responding to temperature differences, on photo-electric cells, or by matching temperatures.

The output of the sun is constant as far as present measurements show and is, at the limit of the earth's atmosphere and at right angles to the sun's ray, 2.000 cal/cm^2/min. This is termed the Solar Constant. Langley is a name given to the unit, cal/cm^2.

2.3 Solar radiation. The revolution of the earth about the sun and its rotation on an axis of almost constant direction in space gives the changing elevation of the sun and latitudinal variations in length of day and solar radiation. For example, at the summer solstice, lat. 0° has 12 hr daylight, lat. 31° 14 hr, lat. 49° 16 hr, and lat. 66·5° 24 hr. Table 2.2 gives the length of day in terms of latitude and time of year.

Table 2.2. *Length of day—sunrise to sunset (northern N, southern S) hemisphere.*

Time of year/ latitude N or S (deg.)	*N Jly.17 S Jan.15*	*Aug.19 Feb.15*	*Sept.17 Mar. 15*	*Oct.18 Apr. 15*	*Nov.17 May.15*	*Dec.16 Jne. 15*	*Jan.13 Jly. 15*	*Feb.11 Aug.15*	*March.13 Sept. 15*	*Apr.12 Oct. 15*	*May 13 Nov 15*	*June 14 Dec. 15*
0	12·1	12·1	12·1	12·1	12·1	12·1	12·1	12·1	12·1	12·1	12·1	12·1
10	12·6	12·4	12·2	11·9	11·7	11·5	11·6	11·8	12·0	12·3	12·6	12·7
20	13·2	12·7	12·2	11·7	11·2	10·9	11·0	11·4	12·0	12·6	13·1	13·3
30	13·9	13·1	12·3	11·4	10·6	10·2	10·4	11·0	11·9	12·8	13·6	14·1
40	14·6	13·6	12·4	11·1	9·9	9·3	9·6	10·5	11·8	13·2	14·3	15·0
50	15·9	14·2	12·5	10·6	9·0	8·1	8·5	9·9	11·7	13·6	15·3	16·3
60	18·0	15·3	12·7	10·0	7·4	5·9	6·6	8·7	11·5	14·3	17·0	18·7
70	24·0	17·6	13·2	8·8	3·6	0	0	6·7	11·3	15·2	21·3	24·0
80	24·0	24·0	14·6	4·2	0	0	0	0	11·2	20·2	24·0	24·0
90	24·0	24·0	24·0	0	0	0	0	0	0	24·0	24·0	24·0

The variation of radiation with latitude can be illustrated by Fig. 2.2 where summer is the period between the spring, autumn equinoxes and winter the remainder of the year. The percentage scale on the right has 100% as the total solar radiation at the equator. Latitudinal and seasonal variations can be easily deduced. The strong influence of latitude and season on radiation, and so weather and climate, is immediately obvious.

The seasonal and latitudinal distribution of the solar radiation received at the top of the atmosphere in cal/cm^2/day, or Langleys/day, is shown in Fig. 2.3. This figure may be used to estimate the solar radiation at a required latitude and time of year.

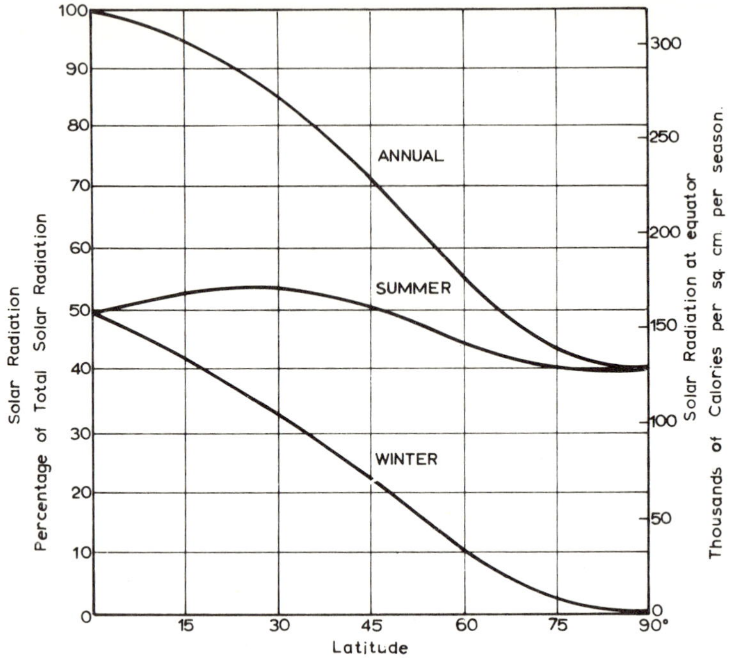

Fig. 2.2. Latitudinal variations of annual, summer and winter solar radiation at top of the atmosphere.

Only a portion of the incoming solar beam is received at the earth's surface as direct radiation, the remainder is reflected, absorbed or scattered by the earth's atmosphere and surface. The proportion received depends upon the nature of the earth's surface and the composition and extent of the atmosphere.

Table 2.3 gives the percentage disposition of inward solar radiation under different weather conditions.

Table 2.3. *Percentage disposal of inward solar radiation.*

Disposition of solar radiation	*Type of day*		
	Clear	*Overcast*	*Cloudy* (52%)
Reflected and scattered back to space	21	75	42
Absorbed by water vapour	11	3 (water vapour and gases)	11
Absorbed by gases	2		4
Direct and scattered reaching earth's surface	66	22	43

The proportion of radiation reflected from a surface is known as its albedo. Table 2.2 shows that the overall albedo of the earth on a clear day is 0·21, on an overcast day 0·75, and an average figure is 0·42. The presence or absence of cloud obviously

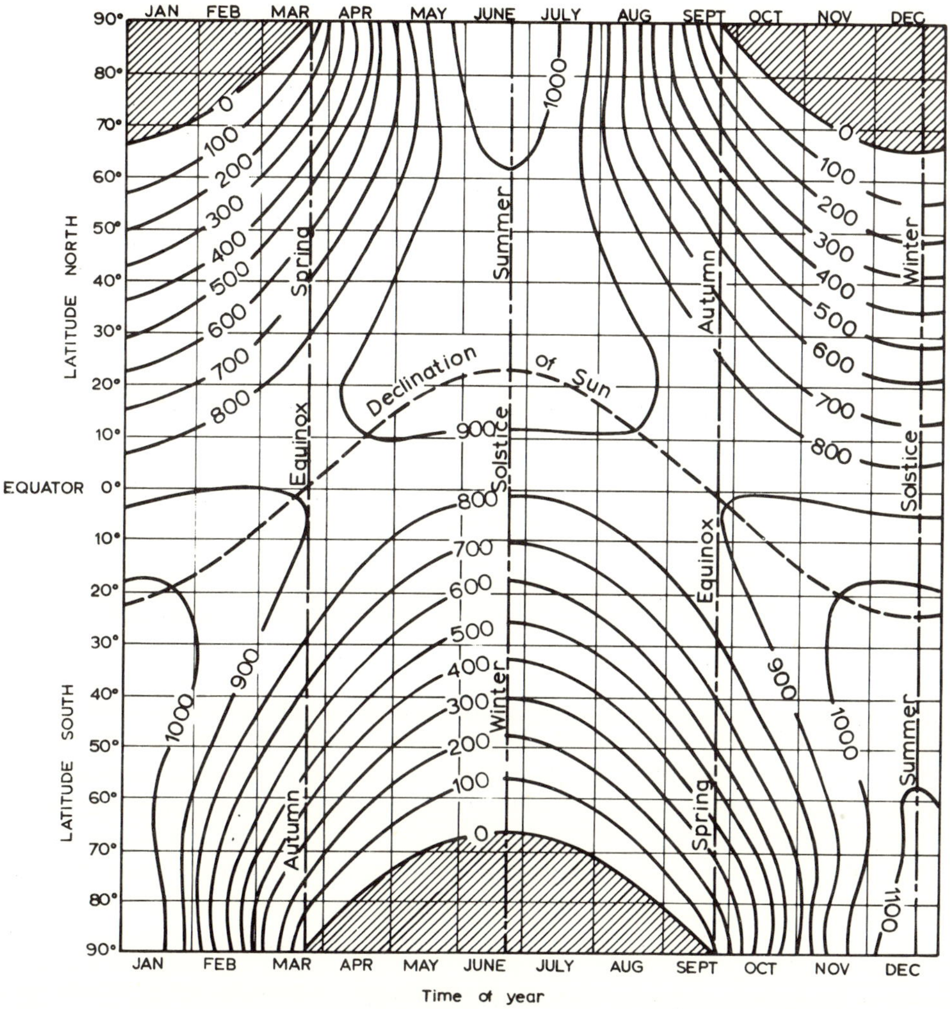

Fig. 2.3. Total daily solar radiation at the top of the atmosphere.

decides the albedo of the earth on a particular day.

Radiation is also reflected from different parts of the earth's surface which have the following albedos.

Dark forest	0·05
Oak woodland	0·18
Pine forest	0·14
Grass and vegetables	0·26
Bare ground	0·10–0·20
Snow	0·46–0·81, dependent on condition.
Water	0·04–0·39, dependent on angle of incidence.

The amount of radiation absorbed by the atmosphere depends upon its composition, the wavelength of the incoming beam and its length of passage through the air.

Water vapour and carbon dioxide, highly variable components of the atmosphere, absorb varying quantities of radiation depending on the amounts present on the day in question. This explains the differences in 'burning power' of the sun on humid and dry days. Longer wavelengths are also absorbed more easily than the shorter ones. This selective absorption is illustrated by the so called 'greenhouse effect' in which short inward waves penetrate the atmosphere but long outward waves are absorbed.

The length of passage of the solar beam through the atmosphere depends upon the elevation of the sun with maximum absorption occurring in the morning and evening, and during the winter season. The air mass path is a measure of the absorption. The water vapour content of the atmosphere or its precipitable water may be used to assess the air mass path and the extent of absorption by water vapour. Thus the position of the earth with respect to the sun not only determines the amount of radiation received at the outer limit of the atmosphere, but also the length of air mass path and so the amount which is absorbed. Latitude and season determine both radiation intake and radiation absorption.

The presence of particles in the atmosphere, smaller than the wavelength of light, causes a scattering of the solar beam. Blue light is scattered most easily, giving the blue colouring to the sky. At sunset and at dawn when the optical path is longest, blue light may be scattered outwards, so that the reddish portions of the incoming light are received directly. On a cloudy day approximately 9% of the solar radiation is scattered into space, and 16% to the earth's surface.

2.4 Terrestrial, or back radiation. The absorption of solar radiation raises the temperature of the earth's surface and its atmosphere. The temperature and energy distribution is complicated and there is a constant interchange of energy between the earth's surface, the atmosphere, and space. However, the earth approaches an average temperature of 294 K°, radiates as a black body, and gives out long electromagnetic waves in the range 4–80μ with a maximum at 10μ. Water vapour in the atmosphere is a strong absorber of long waves, particularly between 5·5 and 7μ and above 27μ, but transmits waves between 8 and 13μ.

By shielding the underneath side, a radiometer can be used to measure the solar radiation at the earth's surface. It will measure terrestrial radiation if the upper side is shielded. If neither side is shielded, the net radiation will be measured, or the solar radiation minus the terrestrial radiation. In a given period this may be either positive or negative.

2.5 The heat balance of the atmosphere. The earth's surface and atmosphere receives solar radiation as short waves and gives back terrestrial radiation as long waves.

There is an excess of incoming radiation at low latitudes and surplus of outgoing in high latitudes. This means an uneven temperature distribution which would be more pronounced but for the transfer of energy over the earth and through the atmosphere in complicated ways. Evaporation, condensation, wind systems, ocean currents and weather phenomena are all processes which re-allocate the energy. At latitude 40° approximately 10^{24} calories are transferred toward the pole each day.

The amounts and balance of solar and terrestrial radiation at a particular place vary from day to day, season to season, and from year to year. However, over the earth as a whole and in a reasonably long period of time, there is a balance between the incoming and outgoing radiation. A simplified version of the heat balance of the earth is given in Fig. 2.4 and shows the processes by which the energy is partitioned. It is assumed that 100 units of solar radiation are initially received at the outer atmosphere.

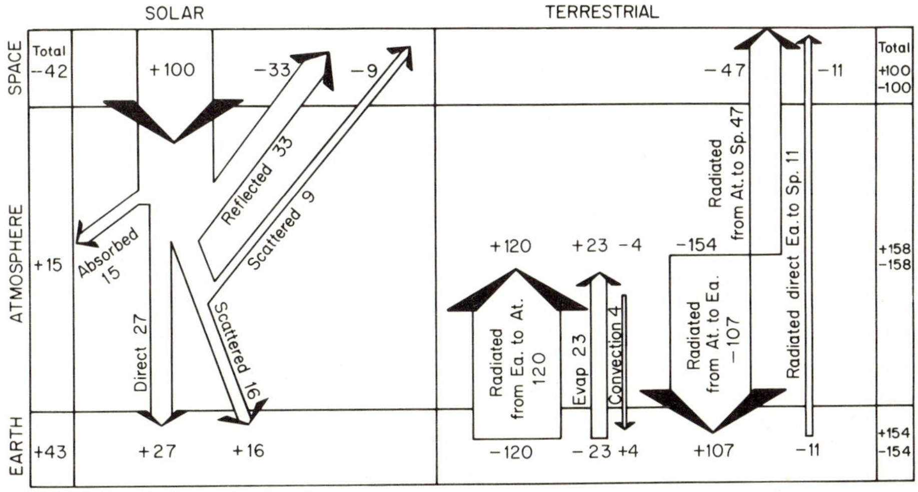

Fig. 2.4. The heat balance of the earth and atmosphere.

Solar radiation from space is disposed as follows:

To space	by reflection from cloud and the earth's surface		33
	by scattering in the atmosphere		9
	Albedo = 0·42	Total	42
To atmosphere	by absorption by water vapour		11
	by absorption by gases etc.		4
		Total	15
To earth's surface	by direct radiation		27
	by scattering in the atmosphere		16
		Total received	43

Terrestrial radiation from the earth's surface is both absorbed by the atmosphere and directly transmitted to space. The atmosphere, in its turn, radiates into space and back to the earth's surface. The overall exchange is shown in Fig. 2.4 and is summarised below.

The long wave or terrestrial radiation moves from

Earth's surface to atmosphere	by l.w. radiation and subsequent absorption	120
	by evaporation	23
Earth's surface to space	by direct l.w. radiation	11

Atmosphere to earth's surface	by l.w. radiation	107
	by convection	4
Atmosphere to space	by l.w. radiation	47

Considering the overall balance

Earth's surface gains	by solar radiation	43
	by convection and l.w. radiation	111
	Total gain	154
Earth's surface loses	to atmosphere by radiation and evaporation	143
	to space by l.w. radiation	11
	Total loss	154
Earth's atmosphere gains	by solar radiation	15
	by evaporation and l.w. radiation from earth's surface	143
	Total gain	158
Earth's atmosphere loses	by l.w. radiation to space	47
	by l.w. radiation to surface	107
	by convection to surface	4
	Total loss	158
Space gains	by reflected and scattered solar radiation	42
	by l.w. radiation from surface	11
	by l.w. radiation from atmosphere	47
	Total gain	100
Space loses	by direct solar radiation	100
	Total loss	100

A global thermal balance occurs over a certain period of time, usually about a year. However, over shorter or longer periods the atmosphere and earth's surface can store solar energy, directly by increases in temperature; for example, local heating in summer, or warming up over geological eras. Energy is stored indirectly as chemical energy through plant activity (photosynthesis) or chemical changes. A one-degree rise in the temperature of the ocean also constitutes a large store of energy. The earth itself has a store of chemical and heat energy, an example of which is volcanic activity.

These heat sources, or storages, are not normally of interest in short term changes, and are neglected as small in relation to the daily receipt of solar energy, but they may accumulate to become of interest and importance when longer term climatic changes, or periodicities, are being considered.

2.6 Computing solar radiation (R_c) at earth's surface. From the solar constant and position of the earth in its orbit, the solar radiation at the limit of the earth's atmosphere (R_a) may be computed. It is given in Fig. 2.3 and Table 2.4 in terms of latitude and

Table 2.4. *Total daily solar radiation at top of the atmosphere* (cal/cm^2/day). *Solar constant* 2·00 cal/cm^2/min.

Time/ Lat. (deg.)	*Mar.* 21	*Apr.* 13	*May* 6	*May* 29	*June* 22	*July* 15	*Aug.* 8	*Aug.* 31	*Sept.* 23	*Oct.* 16	*Nov.* 8	*Nov.* 30	*Dec.* 22	*Jan.* 13	*Feb.* 4	*Feb.* 26
N 90		436	795	1029	1109	1024	788	431								
80	160	436	783	1014	1092	1009	777	431	158	7						7
70	316	541	771	967	1042	962	764	535	312	133	25				25	135
60	460	654	833	962	1008	957	825	648	455	281	150	74	50	75	150	284
50	592	754	893	987	1019	983	885	747	585	426	295	210	181	211	298	432
40	707	831	937	1001	1021	996	928	822	697	561	442	358	326	361	447	570
30	798	891	957	996	1004	989	949	882	788	683	581	507	480	509	585	690
20	866	921	951	963	963	958	943	911	856	783	706	646	623	649	712	792
10	908	924	920	907	899	903	913	914	897	860	813	770	755	775	819	870
0	922	899	862	828	814	824	855	889	912	913	898	877	868	881	904	923
S 10	908	849	783	728	708	725	776	838	897	937	954	959	961	964	964	949
20	866	773	680	611	584	608	674	763	856	934	987	1017	1029	1023	997	946
30	798	674	559	479	449	477	554	665	788	903	993	1051	1072	1056	1002	915
40	707	554	425	339	306	338	421	549	697	844	972	1058	1091	1063	982	853
50	592	420	284	199	170	198	282	416	585	765	928	1044	1088	1048	936	775
60	460	277	144	70	48	70	143	274	455	663	865	1113	1077	1022	872	672
70	316	131	24				24	130	312	548	801	1023	1113	1028	809	555
80	166	7						7	158	442	814	1072	1166	1077	820	447
90										442	825	1088	1183	1094	833	447

time of year.

If the properties of the atmosphere are known it is possible to determine the depletion of this solar beam within the atmosphere. A chart due to Elsasser [1] may be used and is described in texts on Meteorology.[2] Alternatively simple empirical formulae using mean cloud amount (C) as a fraction of sky covered are available and can be used.[3]

e.g.
$$R_c = R_a\,(0{\cdot}803 - 0{\cdot}340C - 0{\cdot}458C^2), \tag{2.1}$$

or if n is duration of bright sunlight and N the number of daylight hours,

$$R_c = R_a\left\{0{\cdot}3 + 0{\cdot}5\,\frac{n}{N}\right\}. \tag{2.2}$$

The constants 0·3 and 0·5 may vary slightly with location. Taking into account the albedo or reflectibility (r) of a surface, the amount of radiation which penetrates the surface is

$$R_c\,(1-r) = R_a\,(1-r)\left(0{\cdot}3 + 0{\cdot}5\,\frac{n}{N}\right). \tag{2.3}$$

2.7 Computing terrestrial radiation from the earth's surface (R_b). The earth may be assumed as radiating as a black body at its mean air temperature. The outward passage of this long wave radiation is reduced by cloud cover, and to a lesser degree by water vapour.

The following expression gives the relationship

$$R_b = \sigma\,T_a^4\,(0{\cdot}56 - 0{\cdot}09\sqrt{(e_d)})\left(0{\cdot}1 + 0{\cdot}9\,\frac{n}{N}\right). \tag{2.4}$$

here σ is the Stefan's constant $= 8{\cdot}17 \times 10^{-11}$ cal/cm²/°K^{-4}/min, e_d is in mm, and T_a in degrees absolute.[4] Values of $\sigma\,T_a^4$ for black body temperatures T_a are given by Fig. 2.5 and in Table 2.5.

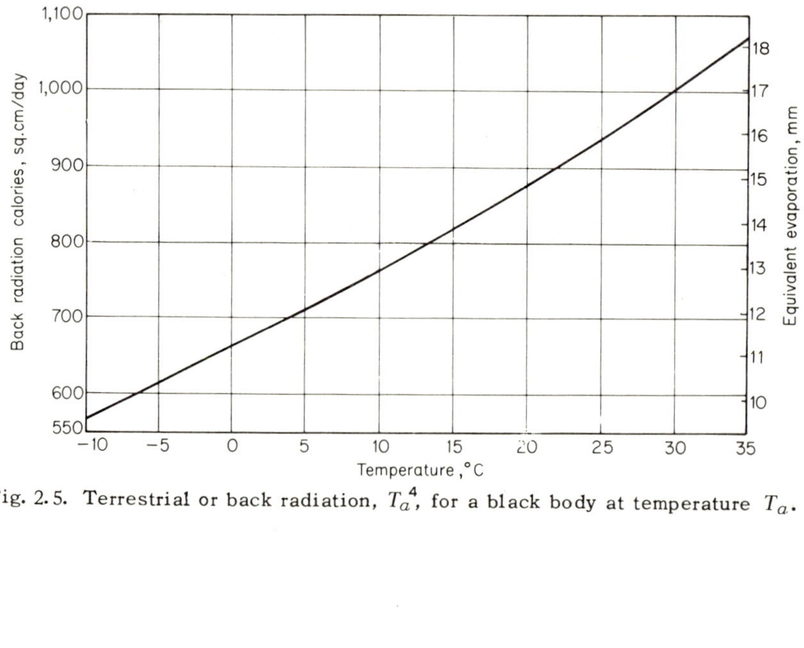

Fig. 2.5. Terrestrial or back radiation, T_a^4, for a black body at temperature T_a.

Table 2.5. *Black body radiation* (cal/cm^2/day) *in terms of temperature* $^{\circ}$K).

Temp.	0	1	2	3	4	5	6	7	8	9
270	622	632	641	650	660	670	680	689	699	709
280	720	730	741	751	762	773	784	794	806	817
290	828	840	851	863	875	887	899	911	923	936
300	949	961	974	987	1000	1013	1027	1040	1054	1068
310	1081	1096	1110	1124	1138	1153	1168	1183	1198	1213
320	1228	1243	1259	1275	1291	1307	1323	1339	1355	1372

2.8 Calculation of net radiation (R_n). The net radiation received by a surface of albedo 'r'

$$R_n = R_c\ (1 - r) - R_b\,. \tag{2.5}$$

The empirical equations given above enables the net radiation received by a surface to be calculated,

i.e. $$R_n = R_a\,(1 - r)\left(0{\cdot}3 + 0{\cdot}5\,\frac{n}{N}\right) - \sigma\,T_a^4\,(0{\cdot}56 - 0{\cdot}09\sqrt{(e_d)})\left(0{\cdot}1 + 0{\cdot}9\,\frac{n}{N}\right).$$

This expression is used in the surface energy approach of computing evaporation from known surfaces using commonly available meteorological data. It is an approximation and is to be used only when actual measurements cannot be made.

A discrepancy of more than 30% between measured and calculated values of net radiation have been found at Davis U.S.A.[5]

REFERENCES

[1] ELSASSER W.M., (1942). Heat transfer by infra radiation in the atmosphere. *Harvard Met. Studies No. 6.*, Harvard University, U.S.

[2] BERRY F.A., BOLLAY E. and BEERS N.R., (1945). *Handbook of Meteorology*, McGraw Hill., N.Y.

[3] BLACK J.N., (1956). Distribution of solar radiation over the earth's surface. *Arch. Met. Geoph, Bio L B. Bd. 7H.2,* p. 165–189.

[4] BRUNT D., (1939). *Physical and Dynamical Meteorology*, Cambridge University Press.

[5] PRUITT W.O., (1960). Correlation of Climatological data with water requirements of crops. *Univ. Calif. Agric. Eng. Dept. Annual Report* (1959–60).

OTHER REFERENCES

BYERS H.R., (1944). *General Meteorology*, McGraw Hill, N.Y.

SLATYER R.O. and McILROY I.C., (1961). *Practical Microclimatology*, U.N.E.S.C.O.

Smithsonian Meteorological Tables (1958). Smithsonian Institute, Washington., D.C.

SELLERS W.D., (1965). *Physical Climatology*, Chicago Press.

CHAPTER THREE

The circulation of the atmosphere and climate

3.1 The factors involved in climate. The climate of a locality is primarily determined by the net radiation which it receives and the atmospheric circulation. In turn, the net radiation received is dependent upon the fixed conditions of latitude and altitude as well as the slope of the land surface and its nature. The resulting variations in radiation over the earth's surface sets up an atmospheric circulation which transports heat, moisture, and solids throughout the atmosphere. The intensity of this circulation and the amount of transport, varies from year to year and affects the radiation balance. The two are interdependent.

The overall radiation balance on a homogeneous earth is considered first, and then the global atmospheric circulation which satisfies such conditions. The modification to this circulation by physiography is then examined.

In this process, the atmosphere is considered as a turbulent fluid which is moving over a rough, rotating spheroid subjected to strong thermal influences. The laws of physics and fluid mechanics apply but, the problems of experimentally determining the atmospheric state at certain times, setting up satisfactory theories and then solving them, are extremely complicated. The difficulties and controversies which arise in the study of past fluctuations of climate and in forecasting weather, are sufficient evidence of the complexity of the problem and the uncertain and unsolved state of knowledge. However, with many simplifications, it is possible to postulate the type of circulation and the major wind and pressure systems which might be expected.

3.2 Air movements on a homogeneous globe. If a homogeneous and non-rotating globe were uniformly heated, convective activity would occur within the atmosphere to the height of the tropopause. No hemispherical scale currents would be expected and the principal storm system would be of convective or thunderstorm types. Figure 3.1 shows such a circulation, where the extent of the atmosphere is, as elsewhere, very much exaggerated for the sake of clarity.

However, if the globe were unevenly heated, with the major intake at the equator and the least at the poles, convection might still be expected but the tropopause would vary in height from the equator to the poles. There would be ascending air at the equator and descending air at the poles. This would result in the movement of warm air aloft from the equator to the poles and cooler air on the surface moving from the poles to the equator.

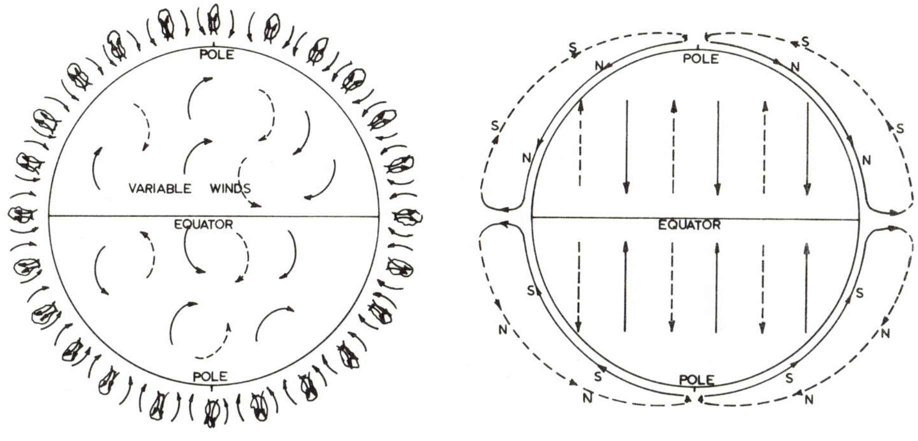

Fig. 3. 1. Air movement on a non-rotating planet.

The overall circulation shown in Fig. 3.1 would be:

	Surface wind	Wind aloft
Northern hemisphere	North	South
Southern hemisphere	South	North

This wind system, if supplied each day with solar energy would lose energy by the frictional losses at the earth's surface and the escape of heat as terrestrial radiation, and the balance would increase wind velocities. Ultimately, a circulation would develop to give a heat balance over the earth.

The rotation of the earth from west to east has the effect of changing the direction of the winds in the following way. A ring of air at the equator and at rest relative to the earth's surface, is spinning about the polar axis at the same speed as the earth's surface at the equator. If the ring of air is forced to move towards the poles its radius will be reduced and, for its momentum to remain the same, its speed will increase from west to east. In other words, it tends to become a westerly. This is known as the Coriolis effect [1] and increases with higher latitudes and in fact is proportional to $v\omega \sin\phi$ where

v is the horizontal wind velocity
ϕ is the latitude
ω is the angular velocity of the earth.

If it were not for the frictional effects of the earth's surface and the lack of symmetry of heat sources and sinks, very strong winds would result. Similarly, if a ring of air is moved from the pole to the equator, the earth will accelerate with respect to the air which will appear to be diverted to the east.

The circulation on a homogeneous earth heated at the equator and rotating would be:

	Surface wind	Wind aloft
Northern hemisphere	North East	South West
Southern hemisphere	South East	North West

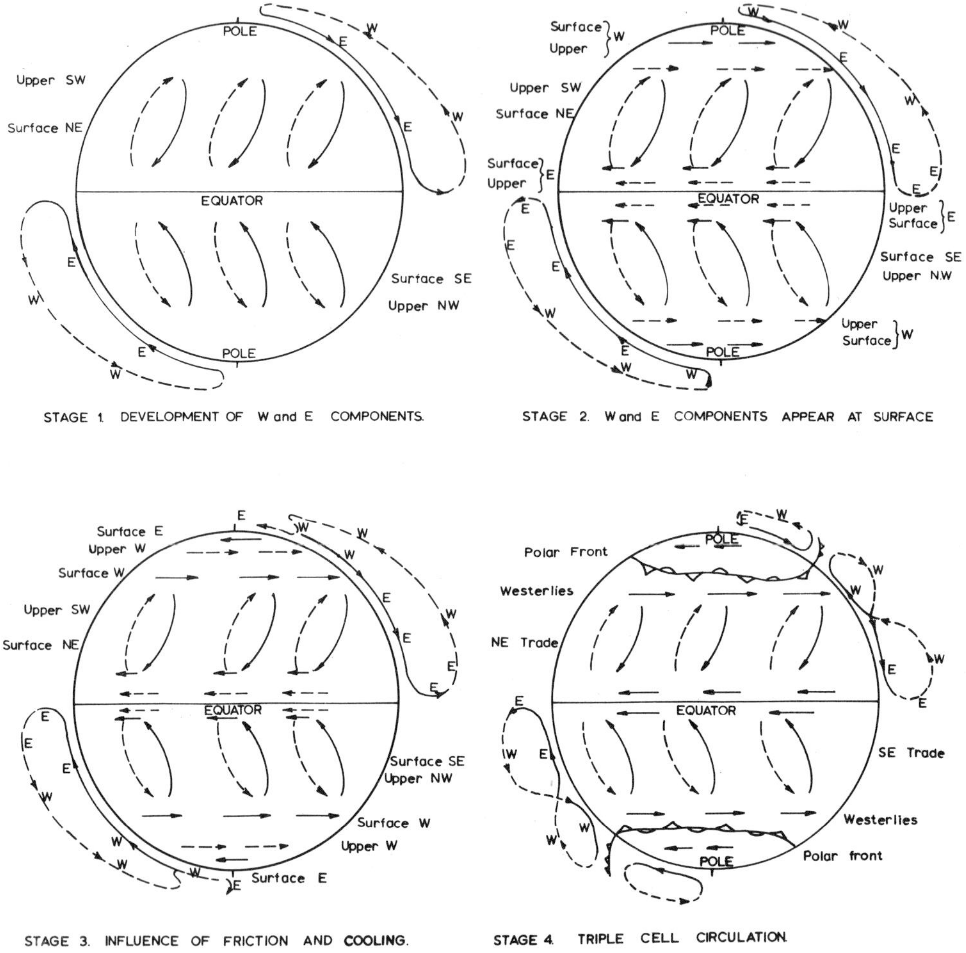

Fig. 3.2. Air movement on a rotating planet.

Such a circulation, illustrated in Fig. 3.2, would result in the surface easterly retarding the earth's rotation. A surface westerly would accelerate it. As the earth's rotation remains constant it follows that the retarding effect of a belt of easterlies is to be offset by the accelerating effect of westerlies in the surface layers. Thus the single cell postulated would split up into a number of cells, at least three, in each hemisphere. On theoretical grounds it has been shown that at least 3 such cells will always exist and this has been found to be true in practice.

The existence of a Corioli's force which occurs when a ring of air moves faster or slower than the earth's surface, involves the building up of a pressure gradient. For example, if a ring of air parallel to a latitude circle is moving faster than that circle the excess centrifugal force tends to throw the ring of air away from the axis of the earth. This results in a force to the south in the northern hemisphere and to the north in the southern hemisphere. To maintain equilibrium, pressure forces build up to counteract these Corioli's forces as follows:

	West wind	East wind
Northern hemisphere	Corioli's force to S compensated by pressure gradient to N and higher pressure to S	Corioli's force to N compensated by pressure gradient to S and higher pressure to N
Southern hemisphere	Corioli's force to N compensated by pressure gradient to S and higher pressure to N	Corioli's force to S compensated by pressure gradient to N and higher pressure to S

This leads to Buys–Ballot's Law which states that:

In the Northern hemisphere an observer with his back to the wind will have a higher pressure to his right.

In the Southern hemisphere an observer with his back to the wind will have a higher pressure to his left.

It follows that westerly surface winds in the northern hemisphere have a high pressure to the right and in the southern hemisphere to the left. This is a mutual relationship between wind, Corioli's force and pressure gradients.

The atmosphere under discussion is axially symmetrical and individual parcels of air, except for losses due to friction, should maintain the same absolute momentum. To do this at the equator and poles the overall north-south movements should create westerly winds near the ground at the poles and easterly winds aloft over the equator. This is shown in Fig. 3.2, stage 3.

Buys–Ballot's law also demands a high pressure area between these latitudes.

The westerly near the poles will lose momentum because of frictional contact with the ground, so that this shallow portion of the belt next to earth's surface seeks its way towards the equator. As air continues to cool and sink near the poles, the retarded westerly winds escape from aloft some distance from the pole. Thus a cellular state sets up similar to Fig. 3.2, stage 4, to meet the mentioned requirements.

3.3 The thermal processes in the atmosphere. The triple cell circulation also satisfies the thermal processes which are taking place in the atmosphere. The air which ascends at the equator proceeds polewards at upper levels, loses heat fairly quickly and descends to reach the ground about latitude 30°. Here one branch goes equator-wards. Another branch of descended air goes polewards along the surface to meet the arctic air from the pole. The forced ascent requires heat which is derived from the latent heat of the ascending air.

The cells from the equator to latitude 30° and from the poles to about latitude 60° adjust the heat balance by carrying heat from warm sources to cold. They are termed the trade wind cell and the polar front cell respectively. The middle cell is frictionally driven by the two direct cells.

3.4 Modification of the ideal circulation. The triple cell circulation satisfies the dynamical and thermal requirements on a homogeneous rotating planet heated by the sun. Lack of homogeneity exists on the earth with the presence of the continents and oceans. This, allied with physiography, gives differential heating and cooling, and a widely varying thermal distribution and balance. This modifies the ideal postulated circulation.

However, maps of the mean wind and pressure systems of the earth show the main features of the three cells with the pattern better preserved in the southern hemisphere where the land area is 19·2% compared with the northern hemisphere where the land area is 39·2%.

The following features of the triple cell circulation are observed on the earth's surface and are significant:

(*a*) The north east and south east trade winds.

(*b*) The prevailing westerlies of middle latitudes.

(*c*) Easterly winds aloft at the tropics and westerly aloft elsewhere.

(*d*) The doldrums area of lower pressure, strong vertical motion, instability, high temperatures, convection, and heavy precipitation.

(*e*) The high pressure belts about latitude 30° where air is usually descending and is dry and hot. Climates are usually arid here and most of the world's deserts are in these latitudes.

(*f*) The polar front regions of low pressure, strong vertical motions, frequent precipitation with contrasting air masses and temperatures.

(*g*) The polar high pressure areas with descending cold dry air and polar deserts.

The pressure systems associated with the triple cell circulation and the known surface winds are shown in Fig. 3.3.

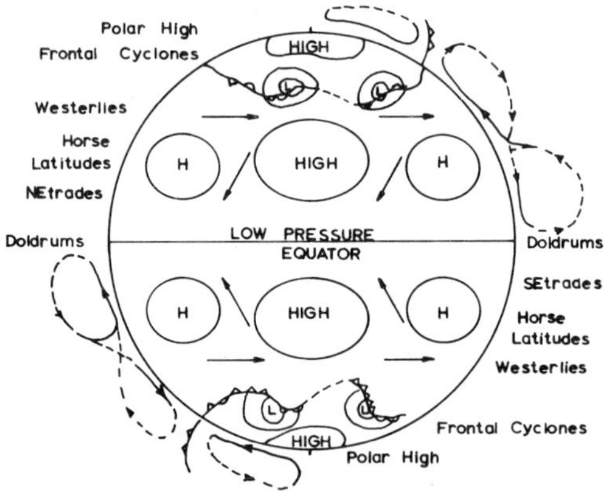

Fig. 3.3. Pressure systems and the triple cell circulation.

3.5 Seasonal movements of the triple cells. The biannual passage of the thermal equator from tropic to tropic gives a similar displacement north and south of the triple cell circulation and explains the seasonal incidence of the prevailing winds and weather systems. The break of the season in the winter rain area, and the monsoons in the tropics, are associated with this annual movement.

In particular years, the thermal effects may be so redistributed that the triple cell circulation may be of greater or less intensity or, it may go further north or south than its usual position. With larger oscillations there can be a more frequent and complete exchange of air between the direct cells and flood years may result. The deep penetration of the monsoons into such countries as India and Australia may occur in these years or cold and warm fronts may affect lower latitudes than they usually do.

Similarly, with a smaller oscillation of the thermal equator, a strong circulation may build up with intense high pressure areas in the horse latitudes. These may be associated with dry periods and slight extensions of monsoonal or cold frontal rains into these latitudes.

3.6 Trajectory of air on a homogeneous planet. If the triple cell circulation is accepted, the trajectories of air may be postulated and are an involved spiral path.

In the northern (southern) hemisphere the NE (SE) trade wind generally rises at the equator and, at height, a portion drifts easterly and then towards the pole. In this passage it acquires a westerly component and, as a W to SW (W to NW) aloft, it cools, with a portion descending near latitude 30° in the subtropical high pressure area, becoming stable, dry, and warmer in the process. At the surface, this air moves out from the high pressure system either, as a W to SW (W to NW) toward the poles, or to become a NE (SE) wind, as it moves towards the equator. Each branch is modified by the surfaces over which it moves, and arrives at higher latitudes or the equator, with vastly different stability, moisture, and temperature characteristics.

The poleward surface portion increases its westerly component to interact at the polar front with cold polar air. Forced ascent occurs with some of the air aloft returning to latitude 30° with much of its westerly component velocity although this does decrease.

The polar cell feeds surface cold E to NE (E to SE) air into the polar front with vertical ascent and some returns aloft to the pole, and in the process acquires a westerly component.

In the latitudes of the polar fronts, surges of cold air from the poles, and warm air from the subtropics, may extend S and N (N and S) to give a vigorous exchange of air, water vapour, heat, and energy between the latitudes.

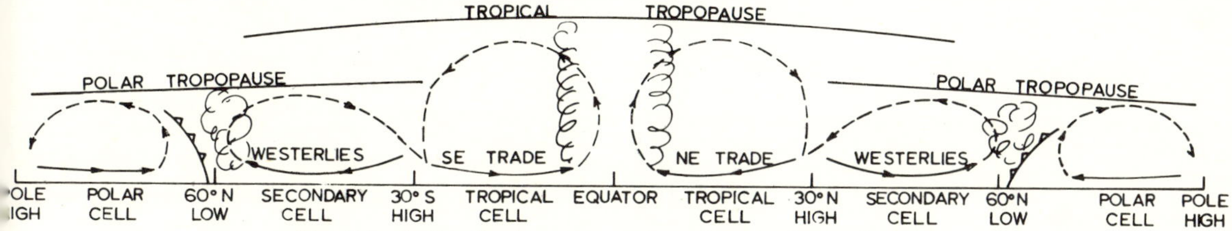

Fig. 3.4. Latitudinal cross section of general circulation.

The turbulence and mixing in the tropics also tend to destroy the identity between air masses and assist the transfer of energy.

A parcel of air would therefore have a contorted, three-dimensional spiral track with many alternative paths. It may undergo many changes of properties and speed of movement en route and the thermal effects of sea and land further complicate the air movement for a non homogeneous planet. When these are added to vertical movements induced by convection, topography, and storms, a very complicated path will result. The triple cell circulation proposed is an idealised one indicating the major movements of air masses over the earth's surface.

3.7 Effect of land and sea–non homogeneity. The idealised general circulation is distorted by the presence of the land and water masses. The land with its varied nature, proportions, shape, and height has the greater range of effects and climates.

Radiation is irregularly absorbed over land surfaces because albedos vary from low values in dark forests and soils to high values upon snow fields. This radiation also passes into a comparative shallow layer which is generally of low conductivity and specific heat. There are, consequently, high temperature rises associated with insolation during the day and marked cooling with strong terrestrial radiation, at night.

Water surfaces, on the other hand, have a low, fairly constant albedo, and radiation penetrates to considerable depths, up to 1500 ft. Part of the radiation is lost to evaporation, in a more constant fashion than over a land surface and part is accumulated in other sectors of the ocean through currents. Water has a high specific heat also, so that overall temperature rises and falls are much less.

An approximate daily range of land surface temperatures is 8–10°F, (4–6°C) on the coast and 15°F, (9–10°C) inland compared with 2–3°F, (1–2°C) for the sea surface. Seasonal ranges are also greater for land areas which are hotter in summer and colder in winter than corresponding sea latitudes.

A climatological atlas shows this temperature effect during summer when the sea level isotherms, or lines of equal temperature, over the oceans bulge towards the equator but, over the continents, bulge towards the poles. In winter time, the oceans are warmer than the land so that the sea level isotherms bulge towards the poles over the ocean, and towards the equator over the continents.

The temperature effects, arising from non homogeneity of the earth's surface, distort the triple cell pattern and introduce many local winds which are thus mainly thermal in origin.

3.8 The tropical monsoon. Land masses within the tropics are effective, by reason of differential heating, in shifting the thermal equator further north or south than over water surfaces in the same latitudes.

The triple cell shifts with the thermal equator and, at the solstice, one trade wind cell is close to the equator and air from it passes across the equator towards the thermal equator in the opposite hemisphere. This thermal equator is displaced furthest from the equator over the continents. The easterly component of the diverted trade wind is replaced by a westerly through these thermal influences to give the tropical summer monsoon. It appears as a south westerly in the northern hemisphere from the diverted SE trade and, as a north westerly in the southern hemisphere from the NE trade.

Thus, the seasonal shift of the general circulation and the thermal heating of the continents give the surge of the monsoon which, in summer, "breaks" over land masses but not in oceanic areas.

3.9 Irregularity of the Earth's surface or topography. The varied land forms on the earth's surface affect the general circulation by introducing additional frictional forces, modifying the thermal pattern, and disturbing the wind flow.

Over the land area the free air flow is affected by the friction of the surface up to a height of about 2000 ft, but to a lesser extent over the sea.

Land forms and shapes introduce a varied temperature pattern and result in local winds such as land and sea breezes, anabatic and katabatic winds.

Topographical barriers not only introduce thermal effects but also affect the prevailing air masses. Storm systems may be blocked from entering some areas by these physical barriers. Where an orographical barrier intercepts a prevailing wind stream, a precipitation pattern dependent on the physical features is set up. If the air mass is moist and unstable, very heavy precipitation can occur as in the Himalaya Mountains. In this case, the range acts as a trigger mechanism to release the latent instability of the air mass.

In the latitudes of prevailing high pressures, although subsidence or a descending motion of the air may predominate, a mountain chain can force the ascent of air streams so that regular and substantial rainfall may be received. It should be noted that the deserts of these latitudes are frequently flanked by mountain ranges upon which rain falls on their windward sides. With the oscillation north and south of the triple cell, suitable air masses are brought into juxtaposition with such topographic features to produce orographical rainfall.

Between the trade wind and middle latitude cells at least half the precipitation has been estimated to occur from orographical influences.

3.10 Local winds on a non homogeneous planet. Although the comparatively simple thermal pattern of a homogeneous planet is distorted by physiography, the main features of the triple cell circulation can be seen in maps of the mean monthly circulation and pressures. The pattern is particularly clear in the southern hemisphere with its smaller physiographic variations. Daily maps do not show the overall wind systems as clearly because local winds are imposed on the general circulation and other pressure systems are developed because the presence of land and sea areas generate distinctive air masses which interact to give pressure differences and systems. This is discussed later, in more detail, under air masses and polar front theory.

Local winds are common, especially if diverse physiography exists. Thermal contrasts generated by this physiography and the differing nature of surfaces give a temperature pattern varying in time and area. Local air currents developed by these temperature gradients may be in phase with, or opposed to, the basic circulation. Land and sea breezes are examples and are responsible for moderating the climates of coastal areas especially in the tropical and subtropical zones. These thermal winds may be combined vectorially with the pressure system winds to give a resultant of different direction and velocity. For example, a north east sea breeze of 10 knots, developing during the morning where a gradient south east wind of 10 knots was blowing, would result in an easterly wind of 14 knots. The sea breeze commences in the forenoon following the establishment of strong thermal gradients between the land and sea.

It reaches a maximum in the afternoon and dies away of an evening. Under cloudy and low temperature contrasts conditions it is light or non existent. It brings the lower temperatures and higher humidities of the sea surface on to the coastal strip, at times extending inland 40 miles. The land breeze arises at night or early morning when sufficient heat has radiated from the land to give thermal gradients rising towards the sea. With less temperature contrast it is shallow, light, of short duration and extends only a few miles to seawards.

Other thermal winds occur inland where the varied heights and nature of land give sufficient temperature contrasts. Anabatic winds are a gentle upflow of air over hills due to air in the higher parts being warmed more than in the lower areas. The downward flow of air from mountain slopes is known as a katabatic wind and are caused by the more rapid cooling of the higher land. These are discussed further in chapter 6.

3.11 The influences of water, vegetation and human activity on climate. From what has been said about the heat balance and the general circulation, it is possible to deduce the effects of water masses, vegetation and human activity on climate.

As pointed out earlier, the climate of a locality depends upon the net radiation which it receives and the atmospheric circulation. These factors are interdependent and the circulation is the more variable.

Again, a very careful distinction must be made between macroclimate or, the overall circulation measured above the earth's surface and, microclimate, or the conditions near the surface. Much of man's control of his environment and his ability to survive under a wide range of conditions is related to his skill in developing microclimates against nature's extremes. For instance, different types of clothing and coverings have given man boundary layers in which his body temperature of 98·4°F or 37°C may be maintained. Buildings are further examples of this modification of microclimate for human needs. A change of climate usually refers to changes in the freely moving air, modifications of the general circulation or changes in macroclimate. Examples of such changes are the variations in the intensity and duration of the tropical summer monsoons, or the local influences of land or sea breezes upon existing circulations. Such changes are beyond man's capacities to engineer, but have occurred over geological time. The ways in which non homogeneity of the earth's surface or presence of land and sea masses, affect the general circulation have already been discussed, but the presence of smaller water masses such as lakes, reservoirs and inland seas might be now considered. In that they possess a different type of surface and slope than the surrounding land, water surfaces receive and deal with the radiation in different amounts and ways. There will be effects on temperature, humidity, evaporation, winds and precipitation on the immediate shores but will not extend far beyond it unless the amounts of energy are great. This implies large areas of water similar in size to continents and oceans, but is a smaller scale effect. For example, the Great Lakes of North America cover a fraction of the area of the surrounding land masses and, although prevailing winds in winter pick up moisture to give increased cloud, temperature, humidity and precipitation on the windward shores, and in summer the air is stabilised with reverse effects, these modifications are limited to the shores only and the macroclimate is not changed over the continent.

Vegetation also varies the nature and contour of the earth's surface, and so affects the radiation balance in a small way with local effects chiefly within the forest itself, where microclimatological conditions are vastly different from a neighbouring bare soil surface. The macroclimate above the trees is not changed significantly. The fact that vegetation intercepts precipitation and retains it for use in transpiration and growth, rather than allowing it to runoff or evaporate, gives rise to the fallacy that precipitation is increased over an afforested area. The very existence of a forest is usually evidence that reasonable precipitation amounts fall. Within the forest, conditions suitable for new growth are maintained, whereas its denudation permits the accumulation of cold air, frost and strong winds near the soil surface with difficulties in regeneration. As the human race lives within the surface layers of the atmosphere, the establishment of vegetation belts wherever possible modifies climate in these layers to the benefit of the human species. Under conditions of high humidity, cloud, fog, dew and frost, deposition of water will occur on trees and leaves with increases in the total precipitation. This is the process of interception. The amount is usually small and is more effective as a supplement and in improving the quality of a harvest rather than in making a sizable contribution to precipitation.

The presence of human settlement and industrial activity increases the heat output of an area and adds to atmospheric pollution. Changes in the radiation balance may occur with differences in the climatological elements. These effects are again mainly in the microclimatological zone but are not insignificant. The overall change of temperature, increase in fog, cloud and precipitation, as well as reduction of sunshine near big cities is marked but may be considered as occurring in the boundary layer of the atmosphere and earth's surface.

The cultivation of land, establishment of irrigation schemes, and large afforestation programmes affect the albedo of the earth's surface and the radiation balance, but effects are chiefly in the boundary layer and the macroclimate is not affected. Plant and animal life however, is sensitive to their local microclimates which can be modified in small ways and over limited areas. This is the only practical field of climate modification. Changes in the macroclimate and the general circulation involve the expenditure of energy beyond the present capacity of man to supply. Changes of climate in geological time have resulted from variations in the overall radiation balance. Some of the factors which may have caused these are fluctuations in the solar energy intercepted by the earth due to shifts in the relative positions of the sun and planets, variations in the properties of the atmosphere causing modification in radiation received at the earth's surface or escape of terrestrial radiation, and the alterations in the earth's surface properties, including its physiography, nature and disposition.

REFERENCES

[1] LEAVER J.M., (1942). A simplified derivation of the Corioli's Acceleration in two dimensions. *Bull. Am. Met. Soc.* **23**, 298–302.

OTHER REFERENCES

HAURWITZ B. and AUSTIN J.M., (1944). *Climatology*, McGraw Hill, N.Y.

HAURWITZ B., (1941). *Dynamic Meteorology*, McGraw Hill, N.Y.

ROSSBY C.G., (1941). The scientific basis of modern meteorology. "*Climate and Man*" U.S. Dept. of Ag. Year Book, 1941, Government Printing Office, Washington, D.C.

SCORER R.S., (1958) *Natural Aerodynamics*, Pergamon.

SUTTON O.G., (1962) *The Challenge of the Atmosphere*, Hutchinson, London.

TUCKER G.B., (1962) The general circulation of the atmosphere. *Weather*, 17, 320–340.

CHAPTER FOUR

The meteorological elements — Temperature

4.1 Measuring atmospheric processes. In order to evaluate and analyse the processes within the atmosphere, meteorology must measure and describe the elements concerned. These are, pressure, temperature, density, and humidity of the air, and its motion relative to the earth's surface. Derived phenomena such as cloud amount and type, precipitation, and evaporation depths are also measured. The elements are inter-related. The motions and processes of the atmosphere are explained in terms of the changes of pressure, temperature and humidity which are brought about by radiation and its variations.

Because the atmosphere is non-homogeneous and varies in its properties with location, time and height, it is necessary to ensure that representative samples are selected and measured and that places, times, and heights are fully specified. Also, although physical principles such as the laws of gases and thermodynamics are applied to the atmosphere, the problem is complicated as many inter-related variables must be measured to describe the situation and one part of the atmosphere cannot be isolated from the changes occurring in other parts. It is not possible to reduce variables to any extent and many of the physical equations describing weather phenomena are non linear and difficult of solution, even with modern computers. The sequence of events is relevant as well as the extreme or maximum and minimum values. Weather processes demand more complicated models than these of classical physics.

Meteorological elements are usually measured quantitatively, for example, temperatures in degrees Centigrade or Fahrenheit, and rainfall as a depth in inches or millimetres, but qualitative estimates are often given as in descriptions of rain as light, moderate or heavy, and clouds as types such as cirrus or cumulus.

A distinction is made between the continuously varying elements like temperature, humidity and pressure and those which are non continuous or discrete, for example, precipitation.

Most of the quantities measured are scalar, for instance, pressure, rainfall and evaporation, but some are vectors, for example wind, where direction and velocity are significant.

Most of the meteorological elements are related to each other, as in the inverse relationship between temperature and relative humidity. Also, pressure, specific volume and temperature are related in the gas equation $pV = RT$.

The variations of the elements with area can usually be shown on maps by drawing isopleths of the elements. Interpolation from these maps enables estimates to be obtained for most places.

Most meteorological elements vary with time, and can be analysed by statistical and graphical techniques which are well described in texts.[1, 2]

The variation of the elements with height can be established by graphical or statistical methods and permit estimates to be made of the elements at other than the available heights.

4.2 Radiation and its measurement. Radiation has already been described in chapter 2, but it is important here to make the distinction between inward or, solar radiation (R_c), outward or terrestrial radiation (R_b), and the net radiation, (R_n) which may be positive or negative and equals inward minus outward radiation ($R_c - R_b$). Each of these may be measured by radiometers with suitable shielding of the sensitive surfaces. The usual type consists of black surfaces in which are embedded the junctions of thermocouples, and which are maintained at standard conditions by clear polythene dome shelters through which circulate inert gas. These and other instruments mentioned are described in the handbooks of the National Meteorological Services or in texts.[3]

There is a sparse network of radiometers, and information of radiation at a site is not always available. However, empirical equations, relating radiation to the actual sunshine hours (n), total possible sunshine hours (N) or mean cloudiness as a fraction of sky covered (C), have been established. Data on sunshine and cloud is more widely available than radiation readings and a simple calculation gives a reasonable estimate.

The simpler formulae and graphs are given in chapter 2.

4.3 Air temperature measurement. Radiation affects, among other things, the temperature of the air envelope of the earth. At a site, the temperature of the freely moving air is measured as distinct from the temperature of the air near objects. There is some consistency in the macroclimate but not in the microclimate, which can vary widely.

Temperature is measured by using the properties of materials which vary with temperature, e.g. the expansion of a liquid as in a mercury-in-glass thermometers, the differential expansion of two metals in the thermograph, the temperature–electrical resistance variation in the electric resistance thermometer and the thermo-electric effect between dissimilar metals in the thermocouple. All of these are well described in current texts. Mercury-in-glass thermometers are commonly used and their readings are referred to as the dry bulb temperature.

At the same time as the air temperature is measured it is possible to get a measure of the humidity of the air by using a wet bulb thermometer. This is an ordinary, or dry bulb, thermometer whose bulb is covered by a piece of muslin supplied with water. The evaporation of water from the bulb is proportional to the humidity of the air and cools or depresses the wet bulb below the dry bulb temperature. Tables of dry bulb and wet bulb temperatures or, dry bulb and depression of wet bulb temperatures, can give any of the required humidity measurements.

Satisfactory sampling of the air is achieved by permitting air at a given height, about 1 to 1¼ m, to circulate past the thermometers which are shielded from extraneous influences, such as direct radiation from the sun, or reflected radiation from walls or ground. Standard practice is to use a thermometer screen, one type being known as a Stevenson screen. This is a white painted wooden cabinet with a double floor and roof with louvred sides which permit a current of air, approx 1/1½ m per sec under normal conditions, to circulate through it. The floor of the screen is at least 1 m above a cut grass surface in a representative site which is fully exposed to wind and weather. The door faces away from the sun and thermometers are mounted as near as possible to the centre of the screen.

A stronger circulation, above 2·5 m/sec, is obtained by whirling thermometers at a reasonable speed as with sling psychrometers or by forcing air, by means of a fan, past thermometers, protected from extraneous radiation as in the Assman aspirated psychrometer. These instruments should be kept out of direct sunlight and a period of up to 2 min may be needed for the thermometers to reach the free air dry and wet bulb temperatures. A different set of psychrometric tables are used to allow for the higher air velocities. These psychrometers are readily portable and are capable of wide applications, and enable air temperatures and humidities for surveys, construction site works, and industrial purposes to be quickly obtained.

4.4 Variations of temperature with height. The rate of change of temperature with height is termed the temperature lapse rate or vertical temperature gradient. In the troposphere it approximates to a decrease of 3½°F per 1000 ft or 6°C per km. At any particular time and place, the actual or environmental lapse rate differs from this average rate and can be measured by instruments, the most common being the radiosonde. This is a small radio transmitter which emits signals proportional to pressure, temperature and humidity values as it rises through the atmosphere lifted by a helium or hydrogen filled balloon.

The loss or gain of heat from an underlying surface or from the top of a layer of air results in changes in the environmental lapse rate. The lower surface layers of unsaturated air vary greatly and experience high lapse rates during a summer day and inversions or increases of temperature with height at night when the ground and air next to it cools rapidly. These variations are shown in Fig. 4.1.

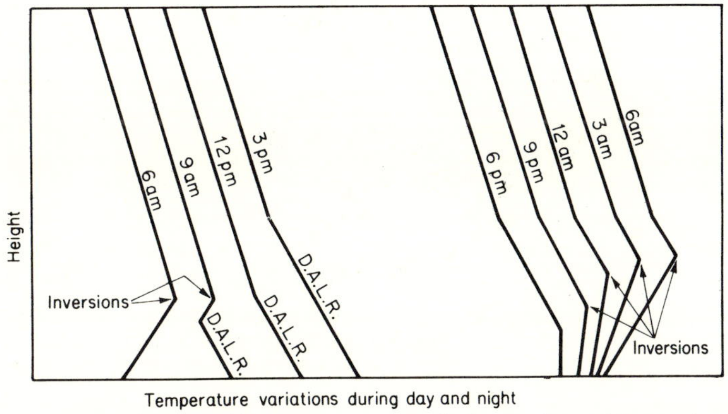

Fig. 4.1. Variation of lapse rates with surface heating and cooling during a day.

(*a*) *The dry adiabatic lapse rate. D.A.L.R.* A process in which a parcel of air neither gains or loses heat is an adiabatic one. This process occurs when air is forced to rise or fall quickly in the atmosphere, say by topographical or frontal action. The air expands as it rises, doing work and taking heat from the air which falls in temperature at a rate for unsaturated air called the dry adiabatic lapse rate (D.A.L.R.) and is 0·54°F per 100 ft or 1°C per 100 m. If, on the other hand, the air descends adiabatically, the higher pressure decreases the volume of air, and the work done is converted into heat, causing the air to rise in temperature, at the dry adiabatic lapse rate. An example of this process is the Föhn wind which occurs on the lee side of mountains. It develops from a wind which rises and condenses its moisture on the windward side of a range. The wind then descends into the valleys on the lee side to become warm and dry. It is known as a Föhn wind in the Alps, but similar winds arise in the same situations elsewhere in the world.

The term 'dynamic' cooling or heating is applied to this adiabatic action as it involves the vertical movement of a mass of air. It is the most effective process available, in nature, for changing the temperature of large bodies of air and is an important stage in the precipitation process.

In nature, the process is not strictly adiabatic because of the interchange of heat at surfaces and between neighbouring air masses but the difference is small and the D.A.L.R. is freely used. Typical natural processes which result in adiabatic cooling or heating are turbulence, movement of air up or down a mountain slope, the upslide or thrusting of one air mass over or into another as in warm and cold fronts, and convergence.

(*b*) *The saturated adiabatic lapse rate. S.A.L.R.* If air is saturated with water vapour, any subsequent cooling may cause condensation with the release of the latent heat of condensation. With saturated air rising in an adiabatic process therefore, the cooling due to expansion is measured against the gain of the latent heat of condensation. This air will cool at a lower rate than dry air and this is termed the saturated or wet adiabatic lapse rate (S.A.L.R.). An average figure for the S.A.L.R. is 3°F per 1000 ft, but varies according to temperature as the amount of heat gained depends upon the quantity of water vapour present and condensed from the air. As greater amounts of water vapour are present at the higher temperatures the S.A.L.R. at such temperatures will be less than at lower temperatures where smaller amounts of water vapour are present and correspondingly smaller amounts of heat of condensation are released.

Thus the S.A.L.R. is smaller in saturated air at higher temperatures. For example, at 68°F the S.A.L.R. is 2·3°F per 1000 ft or at 20°C the S.A.L.R. is 0·426°C per 100 m. S.A.L.R. is greater in saturated air at low temperature. For example, at 32°F the S.A.L.R. is 3·5°F per 1000 ft or at 0°C the S.A.L.R. is 0·645°C per 100 m.

When saturated air approaches very low temperatures (–40°C) only a small amount of water vapour is left in the air and so little latent heat is released on condensation. The S.A.L.R. at such low temperatures therefore (0·950°C per 100 m at –40°C) approaches the D.A.L.R. (1°C per 100 m). The variation of S.A.L.R. with temperature is shown in Fig. 4.2.

(*c*) *The pseudo-adiabatic lapse rate.* A truly adiabatic process carries along with it and cools the products of condensation and also allows for the latent heat of fusion from water to ice. In the atmosphere, water vapour may be supercooled and the proportions of ice, liquid, and vapour are not always in accord with the expectation from physical tables. Also, although truly adiabatic changes may occur in cloud when the water droplets are still in suspension, this is not so when they reach fall-out size and are precipitated from the system. In this case heat is carried away by precipitation. The decrease of temperature with height in saturated air when the condensed water is removed as it is formed is known as the pseudo-adiabatic lapse rate. There is little actual difference in value between the two lapse rates, but the pseudo-adiabatic process is not reversible as the condensation products cannot be replaced. Hence, if an attempt is made to reverse a pseudo-adiabatic process, the air, being unsaturated, warms up at the D.A.L.R.

In nature, condensation products are commonly precipitated and therefore the pseudo-adiabatic processes are applicable. Cloud, which dissipates before it reaches the precipitation stage more nearly follows the adiabatic process.

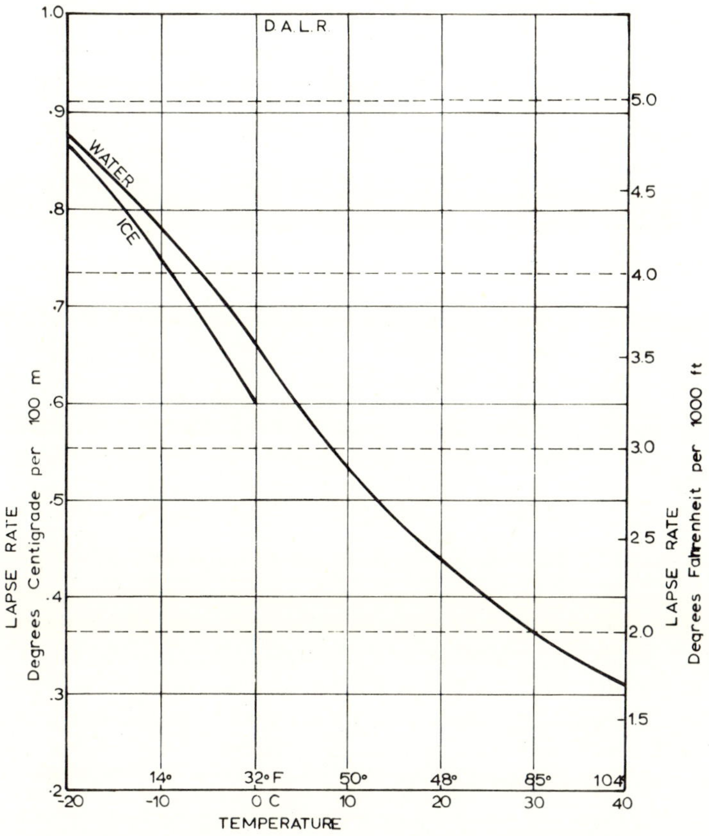

Fig. 4.2. Variation of S.A.L.R. with temperature at 1000 mbs.

(*d*) *The lifting of unsaturated air.* If a parcel of unsaturated air is lifted adiabatically it cools at the D.A.L.R. until it reaches saturation or the dew point. At this condensation level, water droplets may form and become visible as cloud. The cloud base is the condensation level which can be estimated by dividing the difference between the representative surface temperature and the dew point by the D.A.L.R. For example, with air temperature 20°C and dew point 15°C, cloud may be expected at

$$\frac{20 - 15}{1} \times 100\,\text{m, i.e. } 500\,\text{m.}$$

If lifting is continued beyond the condensation level, the air, being saturated, cools at the S.A.L.R. as shown in Fig. 4.3.

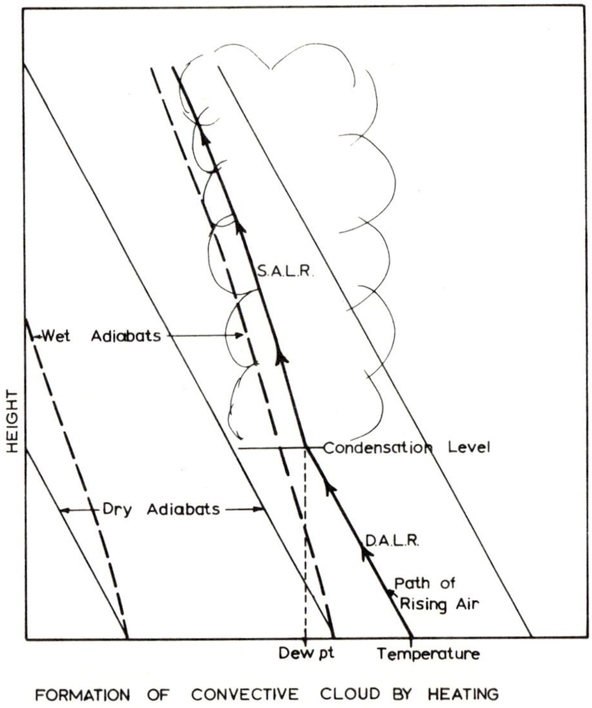

Fig. 4.3. Formation of convective cloud.

4.5 Stability of the atmosphere. The atmosphere is said to be *stable* when a small element displaced through a vertical distance is subjected to forces which tend to restore it to its original position. Air is *unstable* if in its displaced position, it is subjected to forces which tend to displace it further from its original position. The atmosphere is in *neutral* equilibrium if a displaced element is not affected by forces seeking to restore it or displace it further from its original position.

As the displaced element of air takes up the pressure of its surroundings, its density and temperature vary according to the gas laws of physics ($pV = RT$). Density differences and so the stability of the atmosphere can therefore be measured by temperature differences or by lapse rates.

4.6 The stability of unsaturated air. If dry or unsaturated air possesses a lapse rate less than the D.A.L.R. it is stable and, if its lapse rate is greater than the D.A.L.R., it is unstable. This may be illustrated by considering air with a surface temperature of 18°C and an environmental lapse rate (E.L.R.) of 0·8°C/100 m. At 100 m intervals its temperature is as shown in the following Table.

Table 4.1. *Temperatures at 100m intervals with various rates for unsaturated air.*

Level	*Surface*	100 m	200 m	300 m	400 m	*Conditions*
Case 1 E.L.R. 0·8°C/100 m	18	17·2	16·4	15·6	14·8	Stable
Lifted parcel cools at D.A.L.R. 1·0°C/100 m	18	17	16	15	14	D.A.L.R.
Case 2 E.L.R. 1·1°C/100 m	18	16·9	15·8	14·7	13·6	Unstable

If a parcel of unsaturated air is lifted adiabatically from the surface its temperature decreases at the D.A.L.R.

Table 4.1 shows that this parcel at all stages is colder and denser (1°C/100 m) than the environment of Case 1. Because of this temperature and density difference, any lifted parcel of air tends to return to its original position. Case 1 is an example of stable air with a lapse rate less than the D.A.L.R.

In Case 2 where the environmental lapse rate is 1·1°C/100 m, the air is, at all stages, cooler than the lifted parcel which cools at the D.A.L.R. and, being lighter than its environment, continues to rise. Case 2 is unstable air with a lapse rate greater than the D.A.L.R.

4.7 The stability of saturated air. If air is saturated and its environmental lapse rate is less than the S.A.L.R. at that temperature, it is stable, but, if its lapse rate is greater than the S.A.L.R., it is unstable. This is shown in Table 4.2, where an air mass is taken as saturated at 500 m where its temperature and dew point are 13°. In case 3 a lapse rate of approximately 0·4°C/100 m is less than the S.A.L.R. and the air is at all stages warmer than a lifted saturated parcel which cools at the S.A.L.R. and, being denser, tends to return to its original position. The air is stable.

In case 4 the environmental lapse rate is greater than the S.A.L.R. and the air is cooler than the lifted parcel which continues to rise. The air is unstable.

Table 4.2. *Temperatures at 100m intervals with various lapse rates for saturated air.*

Level	500 m	600 m	700 m	800 m	*Condition*
Case 3 E.L.R. 0·4°C/100 m	13	12·6	12·2	11·8	Stable
Lifted parcel cools at S.A.L.R. app. 0·5°C/100 m	13	12·5	12·0	11·5	S.A.L.R.
Case 4 E.L.R. 0·6°C/100 m	13	12·4	11·8	11·2	Unstable

These various cases are shown in Fig. 4.4.

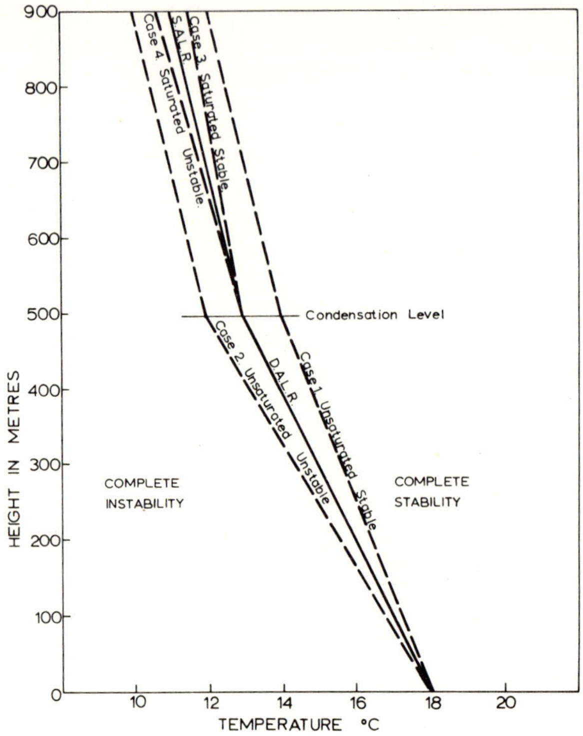

Fig. 4.4. The stability of air of various lapse rates.

4.8 Formation of cloud. When the environmental lapse rate is greater than the D.A.L.R. the air is unstable irrespective of whether the air is saturated or not. Also, if the environmental lapse rate is less than the S.A.L.R., the air is stable for the saturated and unsaturated condition. If, however, the environmental lapse rate lies between the D.A.L.R. and S.A.L.R. then the air is stable while unsaturated but unstable when saturated. The air is said to be conditionally unstable and is stable up to the condensation level where cloud forms. Above this it is unstable and, as saturated air, it continues to rise throughout the depth of conditional instability.

A mountain range has a lifting or ‘trigger action’ on an incident stream of conditionally unstable air and, if lift is sufficient, forms cloud at the condensation level. This cloud continues to grow until the lapse rate is less than the S.A.L.R. and may result in showers or thunderstorms.

4.9 Thermo-dynamic diagrams. Information on the vertical temperature distribution of the atmosphere is obtained regularly from radiosondes and enables meteorologists to make a day-to-day assessment of the stability of the air and the likely behaviour of the air mass when subjected to meteorological processes such as turbulence, fronts, upslide or convergence. This vertical temperature and humidity profile is plotted on a diagram on which the dry and saturated adiabatic lapse rates are shown. The environmental lapse rate can then be compared with the D.A.L.R. and S.A.L.R. and the stability of the air quickly estimated. Such a simple thermodynamic diagram is shown in Fig. 4.5 where case 5 shows an unstable environmental lapse rate dotted while the ascending particle follows the solid line. Cloud grows from condensation level A to cloud top B and the energy available for this growth can be determined from the size of the shaded area between the two curves. In case 6 the air possesses a more stable lapse with less energy released when lifted. Shallow cloud only results when the air mass is lifted to the condensation level.

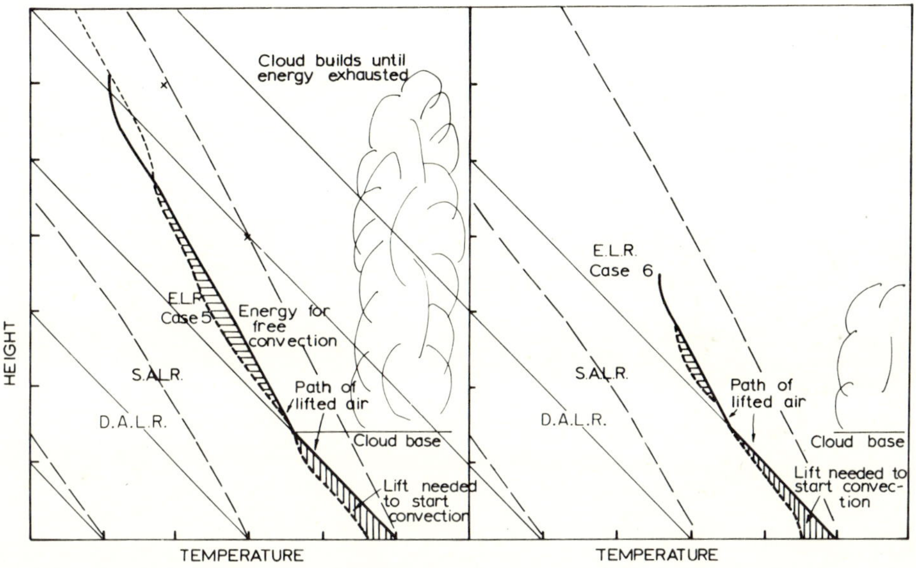

Fig. 4.5. The formation of cloud when an air mass is lifted.

4.10 Variations of temperature with time. Air temperature has a diurnal variation similar to that of radiation but with a lag from one half to two hours depending whether oceanic or continental stations are considered. Thus unless significant weather changes and cloud structures occur, the minimum air temperature occurs shortly after sunrise, and, the maximum in mid afternoon. Local air movement may have an influence especially on the coast where the maximum frequently occurs in the morning, before the onset of the sea breeze which reduces temperatures.

The presence of cloud reduces the amount of inward radiation and prevents the escape of terrestrial radiation. Consequently there is a lower range of temperatures on cloudy days. High atmospheric humidity has a similar effect. The greater daily and annual range away from the coast results from lower humidity and cloud amount inland.

The mean daily temperature is obtained by averaging the available records throughout the 24 hr, one of the most accurate being the mean of the hourly readings. Thermograph traces adjusted to standard thermometer readings and averaged are also satisfactory. The average of the extremes, that is maximum plus minimum divided by two, is a good estimate of the mean temperature and is widely used.

The temperature regime of a locality is well described by the average daily temperatures $\left(\frac{\text{daily max} + \text{min}}{2}\right)$ on a monthly basis and by the extremes in each month. A more detailed picture may be obtained from a frequency analysis, usually on a monthly basis, of the daily maximum, minimum or mean temperatures. The monthly figures may be converted to seasonal or annual analyses.

The curve of monthly air temperatures lags about a month on the receipt of radiation, that is, hottest and coolest, months occur about a month after the solstices or in January and July. There is usually a greater lag at coastal than at inland stations.

In using temperature data, its variation with height must always be considered, as this is much greater than with latitude. A temperature height relationship should be constructed for a locality using available data and the temperatures interpolated at the heights required by the investigations.

REFERENCES

[1] BROOKS C.E.P. and CARRUTHERS N., (1953). *Handbook of Statistical Methods in Meteorology*, H.M.S.O., London.

[2] CONRAD V., (1944). *Methods in Climatology*, Cambridge, Mass.

[3] MIDDLETON W.E.K. and SPILHAUS A.F., (1953). *Meteorological Instruments*, University of Toronto Press, Toronto, Canada.

CHAPTER FIVE

The meteorological elements — Water vapour content of the atmosphere

5.1 Water vapour in the atmosphere. On most occasions water vapour is present in the troposphere but the amount varies widely. A suitable measure of the water content of the air together with the inflow velocity into a storm model gives the available moisture for precipitation. Also, the quantity of water vapour present in the atmosphere affects the absorption and escape of radiation. The various changes of state involve the gain or release of heat energy which is transferred from one area to another and is transformed into other forms of energy such as potential and kinetic. Again, the amount of water vapour contained by the air and its degree of saturation, especially in the micro-climatological zones have important bearings on biological and chemical processes, physical comfort, and the drying of materials.

The oceans, covering four fifths of the earth's surface, absorb a high percentage of the incident radiation and are the source of most atmospheric moisture although transpiration and evaporation from plants, animals, and surfaces make a considerable contribution.

5.2 Atmospheric humidity. Atmospheric humidity is broadly defined as the water vapour content of the atmosphere and is measured in a variety of ways, all of which are inter-dependent. Direct measurement of atmospheric humidity is possible by physical experiments similar to passing a known quantity of air over absorbent material which is increased in weight. The simplest and most convenient method uses wet and dry bulb thermometers and empirical equations which relate these readings to the desired measures of humidity. The more important measures and their inter-relationships follow.

5.3 Vapour pressure. Atmospheric humidity is expressed in terms of vapour pressure or the partial pressure exerted by the water vapour. Its unit of measurement is pressure in millibars, or in the equivalent length of a column of mercury in inches or millimetres.

At a given temperature there is a maximum amount of water vapour which can be held. The space is then saturated at that temperature and the partial pressure exerted by the water vapour is the saturation vapour pressure (e) at that temperature. It is measured in millibars, or in inches or millimetres of mercury.

As the temperature rises more water vapour can be held. Thus the saturation vapour pressure increases with temperature as shown in Fig. 5.1 or Table 5.1. It follows that if the air is not saturated and cooling occurs, then a temperature is reached when it becomes saturated. This temperature is known as the dew or frost point (T_d) as it is marked by condensation and the deposit of liquid water as dew, or ice as frost.

If air is cooled below the dew point without condensation, there is a state of

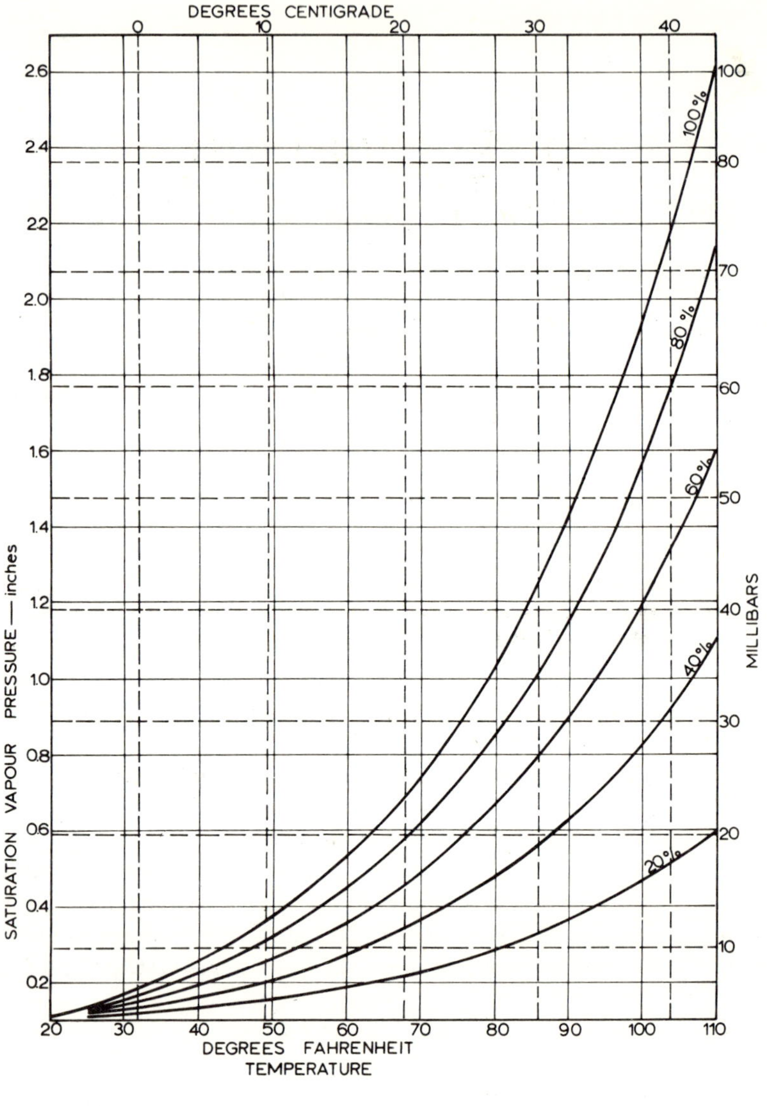

Fig. 5.1. Variation of vapour pressure with temperature at percentages of saturation.

supersaturation, or more water vapour is held than is necessary to saturate the space. Also, water droplets may cool below freezing point without freezing and become supercooled. Conditions of supersaturation and supercooling may occur in the higher atmosphere if condensation nuclei are absent. If suitable nuclei are provided, condensation and freezing occur.

The saturation vapour pressure over ice is less than that over water at the same temperature. Consequently supercooled liquid water or water vapour, in the presence of ice, condenses upon the ice surface. That is, the presence of ice crystals in a cloud leads to the deposit of water upon them, an increase in drop size, and perhaps precipitation.

The vapour pressure of the air (e_d) or saturation vapour pressure at dew point T_d can be measured directly or deduced from empirical equations which use the saturation vapour pressure at wet bulb temperature (e_w), air pressure (p) dry bulb (T_a) and the wet bulb (T_w) temperatures. They are for the whirling psychrometer and air current 2·4 m/sec,

$$e_d = e_w - 0{\cdot}000660p\ (T_a - T_w)\ (1 + 0{\cdot}00115T_w)\ \text{Temperature in }^\circ\text{C} \qquad (5.1)$$

$$e_d = e_w - 0{\cdot}000367p\ (T_a - T_w)\left(1 + \frac{T_w - 32}{1571}\right)\ \text{Temperature in }^\circ\text{F} \qquad (5.2)$$

Similar equations for the Stevenson Screen air current 1–1½ m/sec and degrees Fahrenheit are

$$e_d = e_w - 0{\cdot}000437p\ (T_a - T_w)\left(1 + \frac{T_w - 32}{1098}\right)\ \text{wet bulb above } 32^\circ \qquad (5.3)$$

$$e_d = e_w - 0{\cdot}000437p\ (T_a - T_w)\left(1 + \frac{T_w - 32}{1208{\cdot}2}\right)\ \text{wet bulb below } 32^\circ \qquad (5.4)$$

Here e and p are measured in the same units.

5.4 Saturation deficit or vapour pressure deficit. The difference between the vapour pressure at saturation and the actual vapour pressure at the temperature in question is known as the saturation deficit expressed as $e_a - e_d$ and measures the difference between the existing moisture conditions and the saturated state.

It has applications in measuring the "drying power" of the air and in biological studies.

5.5 Relative humidity. A common measure of the degree of saturation at a given temperature is the relative humidity. This is the percentage ratio of the amount of moisture present to the amount necessary to saturate at that temperature. In the calculation of relative humidity any of the measures of atmosphere humidity may be used, that is vapour pressure, absolute humidity, etc.

$$\text{Relative humidity} = \frac{\text{Moisture present} \times 100}{\text{Amount necessary to saturate}}$$

$$\text{R.H.} = \frac{e_d}{e_a} \times 100 \quad \text{or} \quad \frac{\rho_w}{\rho_{ws}} \times 100. \qquad (5.5)$$

Here ρ_w is the absolute humidity and ρ_{ws} the absolute humidity if air were saturated at given temperature.

As the saturation vapour pressure e_a varies with the temperature T_a as shown in Fig. 5.1 the relative humidity changes according to the amount of moisture present and inversely as the temperature. That is, if the temperature rises and there is no change in moisture content, the relative humidity falls and vice versa. This inverse relationship between temperature and relative humidity is illustrated by the thermo-hygrograph which, in a day of constant moisture content, shows the temperature curve inverse to that of relative humidity.

Relative humidity is significant in the drying out of articles. A relative humidity of less than 70% is usually required if moisture is to pass freely from any media, that is, if it is to dry rapidly. Also, chemical actions, such as rusting are accelerated with high relative humidities, above 65%, especially when these high humidities are associated with high temperatures.

Human comfort is dependent on relative humidity in that it controls the rate of evaporation of perspiration from a body which is then cooled at a corresponding rate. The life, growth, and development of plants, diseases, bacteria, and animals are influenced by relative humidity. For example, moulds grow rapidly above relative humidities of 70%. Also, plants and other organisms frequently seek conditions in the microclimate which give them the high humidities necessary for them to flourish and propagate.

5.6 Absolute humidity (ρ_w). Another measure of humidity is the absolute humidity (ρ_w) defined as the mass of aqueous vapour in a given volume and at a given temperature.

$$\text{i.e.} \quad \rho_w = \frac{\text{Mass of water vapour}}{\text{Volume}} = \frac{M_w}{V} = \frac{\text{g}}{\text{cbm}} = \frac{\text{grains}}{\text{ft}^3}\,. \tag{5.6}$$

It is equivalent to water vapour density or the vapour concentration. For dry air the relationship between pressure (p), density (ρ) and temperature T is given by the gas equation $p = RT\rho$ where $R = 2{\cdot}87$ for dry air when p is in mb, T in K° and ρ in kg/cbm $R = 2{\cdot}87 \times 10^3$ when p is in mb, T in °K and ρ in g/cm³. $R = 2{\cdot}87 \times 10^{-3}$ when p is in mb, T in °K and ρ in g/cbm. Water vapour has a density of $0{\cdot}622$ or approximately five eighths of the dry air density. That is, water vapour density,

$$\rho_w = 0{\cdot}622\,\frac{e_d}{RT} \quad \text{or} \quad \frac{5\,e_d}{8RT}\,. \tag{5.7}$$

If ρ_w is in g/cbm, $R = 2{\cdot}87 \times 10^{-3}$ and then

$$\rho_w = 217\,\frac{e_d}{T}\,. \tag{5.8}$$

Thus the absolute humidity ρ_w is proportional to the vapour pressure at a given temperature.

If p is now the pressure of moist air and e_d is the vapour pressure, then $p - e_d$ is the pressure of the dry air and the density of the mixture

$$= \frac{p - e_d}{RT} + \frac{5e_d}{8RT}$$

$$= \frac{p}{RT}\left(1 - \frac{3e_d}{8p}\right)$$

$$= \frac{p}{RT_v} \quad \text{where } T_v = \frac{T}{1 - \dfrac{3\,e_d}{8p}}\,. \tag{5.9}$$

T_v is termed the virtual temperature and is the equivalent temperature of dry air which has the same density as the damp air at the same pressure. The virtual temperature enables the gas equation to be used with the original gas constant for a mixture of air and water vapour.

Alternatively, a gas constant R_w for moist air may be used and

$$R_w = \frac{R}{1 - \dfrac{3e_d}{8p}} \tag{5.10}$$

Absolute humidity at saturation, ρ_{ws} is the maximum value of the water vapour density at a given temperature. The values of absolute humidities at saturation for various temperatures are given in Table 5.1.

Table 5.1. *Measures of humidity at saturation.*

Temp. °C	*Vap. press.* e, mb	*abs. hum.* ρ_w, g/cbm	*Mix. ratio* r, g/kg	*Spec. hum.* q, g/kg
0	6·108	4·86	3·84	3·82
5	8·719	6·81	5·50	5·47
10	12·27	9·41	7·76	7·70
15	17·04	12·83	10·83	10·72
20	23·37	17·31	14·95	14·73
25	31·67	23·60	20·44	20·04
30	42·43	30·40	27·69	27·02

5.7 Mixing ratio. The mixing ratio (r) of moist air is the ratio of the mass of water vapour (M_w) to the mass of dry air (M_a) with which the water vapour is associated.

i.e.
$$r = \frac{M_w}{M_a} \tag{5.11}$$

The saturation mixing ratio occurs when the air is fully charged with water vapour at that temperature. The variation of saturation mixing ratio with temperature is similar to that of vapour pressure and absolute humidity and some values are given in Table 5.1.

Mixing ratio can also be calculated using the previous gas equations.

$$r = \frac{M_w}{M_a} = \frac{\rho_w}{\rho_a} = \frac{\dfrac{622e_d}{RT}}{\dfrac{p - e_d}{RT}} = \frac{622e_d}{p - e_d}. \tag{5.12}$$

Here r is measured in g/kg, the corresponding fps measure being grains/lb. Mixing ratio is not affected by pressure or temperature changes unless water is added or subtracted by evaporation or condensation. Mixing ratio is a conservative property of an air mass.

5.8 Specific humidity. The specific humidity (q) of moist air is the ratio of the mass of water vapour (M_w) to the mass of moist air ($M_a + M_w$) in which the mass of water vapour is contained.

$$q = \frac{M_w}{M_a + M_w} = 622 \frac{e_d}{p - 0{\cdot}378\,e_d} = 622 \frac{e_d}{p} \text{ approx.} \tag{5.13}$$

Here q is measured in g/kg, the corresponding fps measure being grains/lb.

Saturation values of specific humidity occur when the air is fully charged with water vapour at that temperature. Specific humidity is also a conservative property of an air mass and is not affected by pressure and temperature changes unless water is added or subtracted. Values of saturation mixing ratio and specific humidity are given in Table 5.1 at selected temperatures.

There is little difference, approximately ¼ to 4%, between values of specific humidity and mixing ratio and for most purposes they may be used interchangeably or converted. When measured in g/kg,

$$q = \frac{r}{1 + \frac{r}{1000}} \tag{5.14}$$

5.9 Precipitable water. A convenient measure of the water vapour content of a column of air is the precipitable water (W). This is defined as the total amount of water vapour in a layer of air expressed as a depth of liquid water in millimetres or inches. That is, precipitable water is the depth of water which would result from the condensation of all the water vapour in the column of air. It is of course, impractical for this to occur in nature, and all air masses contain some precipitable water. Precipitable water is considered in more detail in chapter 13 as it has important applications in storm models, maximization, and transposition.

5.10 Measurement of water vapour content of the atmosphere. As mentioned earlier, the water vapour content of the air may be determined by physical experiments but it is more convenient to use wet and dry bulb thermometers, suitably ventilated. The Stevenson screen, with an air current of 1–1½ m/sec and the whirling psychrometer with an air velocity of greater than 2·4 m/sec give slightly different readings of wet bulb temperatures at some ranges of temperature. The empirical formulae and tables for the two conditions are also slightly different but give similar values of water vapour content. The correct formula and tables are to be used for the exposure in question.

Certain materials, including paper, hair, and polythene film vary in size according to their moisture content and come quickly into adjustment with the atmospheric vapour content. They are therefore used to measure humidity with appropriate gears or cams to convert their variation in length to a measure of relative humidity. The hygrograph uses this principle with stretched hair which, under variations of humidity, moves a pen over a rotating drum.

The amount of electric current which can pass across a specially treated glass strip is another measure of the water vapour content of the air and is often used in radiosondes.

5.11 Psychrometric tables, nomographs, slide rules. The formulae quoted enable most of the measures of atmospheric water vapour content to be calculated. Also, graphs enable interpolation, and, if required, can easily be constructed from formulae or tables. In practice, it is quicker and more convenient to use psychrometric tables which give the various measures directly from dry and wet bulb readings. [1–3] There are also nomographs available expressing the relationships in graphical form and greatly facilitate estimations and conversions. [4,5]

Special slide rules and calculators are also available and have advantages in that interpolation is simpler and accuracy is adequate for most purposes.

5.12 Variations of water vapour content with time. At most places on the earth's surface, the annual variation of all measures of water vapour content, except relative humidity, is similar to that of temperature with the highest value occurring in summer and the lowest in winter. However, the diurnal variations of the measures depend upon the degree of mixing of the air in the surface layers (turbulence), the addition or extraction of water by evaporation or deposit (dew, rain etc.), the amount of moisture available for evaporation, (moist soil, water surfaces), and the energy to convert the water into the vapour form, (radiation and heat from air, ground, etc.). Consequently, the surface layers vary in water content because of the deposit of dew or frost at night when surface cooling occurs and because of the upward transfer of moisture by evaporation and mixing following heating during the day. Thus lower values of water vapour content occur in the morning but rise to a maximum before noon as the dew and surface water is evaporated. Convective mixing of the air causes a fall to a minimum in the afternoon, and cessation of this mixing and cooling towards evening again results in increased water vapour content in the surface layers. This is shown in Fig. 5.2.

Relative humidity, because of its inverse dependence on saturation vapour pressure and thus temperature, shows annual and diurnal variations different from the other measures. For example, vapour pressures are low in winter but so are temperatures with lower saturation and vapour pressures, and together these give high relative humidities in winter. In summer, although actual vapour pressures are greater, the saturation vapour pressures are correspondingly higher to give summer relative humidities lower than in the winter season. The diurnal variation of relative humidity is the reverse of the temperature sequence with a maximum at night or early morning and a minimum in the afternoon.

Humidity measurements may be influenced strongly by local features such as inversions, pools of stagnant air, surfaces of water, vegetation, damp soil, air drainage, and local air movements like land and sea breezes.

Because of the numerous factors which affect humidity it is necessary to take every precaution to measure it accurately and to obtain a value which fully represents the water content of the air mass. Dew point readings may be checked by reducing them to a standard level of 1000 mb by using the pseudo-adiabatic lapse rate and by comparing them with other stations, especially those along the air mass's trajectory. Also, readings of humidity should be made over a period of time, say 12 or 24 hr rather than isolated observations which may be subjected to local influences. A dew point considered in these ways is a satisfactory measure of water content and its minimum value over a period is accepted as the persisting dew point over that period of time and as truly representative of the air mass.

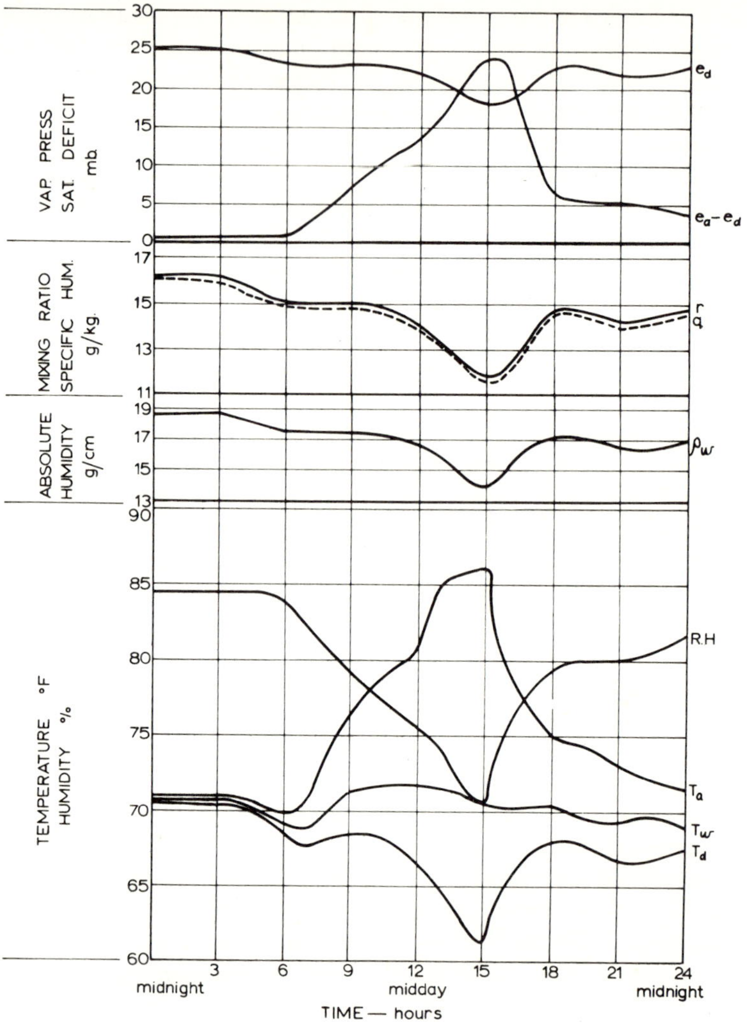

Fig. 5.2. Diurnal variation of temperatures and humidities on a summer day.

5.13 Variation of humidity with height. Most of the atmospheric water vapour originates from the earth's surface and is then distributed into the upper layers, mainly by the processes of turbulence and lifting. As the amount of water vapour held in a given volume depends upon temperature, water vapour contents vary with height in a similar way to temperature. That is, there is a vertical water vapour gradient as well as a vertical temperature gradient or lapse rate with maximum values near the surface. Also, these lower layers display considerable variations in water vapour content and relative humidity both with respect to time and area.

Radiosondes measure the variation of moisture content with height as well as temperature, and enable direct evaluation of the water vapour content of the atmosphere. Where such data is available it should always be used with surface information to obtain a representative value of the moisture content of an air mass. Surface dew points should be consistent with radiosonde readings in the lower layers. Under stagnation conditions, with low wind velocities and inversions, high moisture conditions may exist in the lower layers over local areas and for short periods, e.g. after a heavy thundershower. Such conditions should be recognised so that these high dew points are not taken as indicating high moisture contents throughout the atmosphere or for long periods of time. Thus they do not indicate maximum conditions which could be applied to storms covering larger areas and persisting for some time. This is further considered in chapter 13.

5.14 Variation of humidity with distance. On the coast adjacent to the principal source of water vapour and sea breezes, humidities are highest. In fact, coastal dew points with onshore winds approximate to the seasonal sea temperatures. Inland and away from sources of moisture such as lakes and vegetation, humidities are lower. The differences are greatest in the surface layers because the passage of air inland, with heating from below and turbulence, distributes the water throughout a greater depth of the air mass. However, precipitable water which is evaluated throughout the depth of an air mass, shows a more marked variation with latitude than with distance from the sea.

5.15 Condensation of water vapour in the atmosphere and formation of cloud. Besides the vapour state, water may be present in the atmosphere as a liquid or solid when it is then visible as cloud, mist or fog. For water vapour in the atmosphere to aggregate or condense, the air containing it should be cooled until it is saturated and the latent heat of condensation released. The formation of drops of water and ice proceeds most rapidly when the atmosphere possesses a high water content, it is near saturation and it is subject to a continuing process which is capable of cooling large masses of air. The presence of condensation nuclei facilitates the process.

Air is cooled in a number of ways. It can radiate heat outward, pass heat on to an object in which it is in contact, or mix with colder air. Usually only thin layers of air are cooled in this fashion to form patches of fog, mist or shallow cloud. Large masses of air are most easily cooled by the process of adiabatic expansion described in chapter 4. This involves the rapid reduction of pressure which is best performed by lifting the air. Thus any process which lifts air masses will cause adiabatic expansion and give the possibility of cloud formation and precipitation. When water droplets and ice are formed in this way they are supported by the vertical currents which raised the air mass and led to their formation. Decreases in the velocity field, and the supply of moisture into an area leads to the dissipation of cloud. In reverse, if they increase, there is the continued growth of cloud and, perhaps, the formation of precipitation products.

The processes which give vertical lift to the air masses can be classified as turbulence, uplift over a physical barrier, movement of one air mass over another of different temperature, and convergence. The last is the horizontal movement of more air into a given space than can pass out horizontally and leads to vertical motion. Cloud is often classified according to the processes which created it. For example, large scale turbulence or convection gives cumulus cloud or convective cloud, movement over a mountain barrier gives orographical cloud, and where two air masses are involved there is a complicated pattern of layer and turbulent cloud which is recognisable as a frontal cloud system.

The two obvious characteristics of a cloud are its base or the level at which condensation occurs, and its form which results from the processes which built it. It may be a layer or status variety when the lift is gradual and the air mass stable. The cloud is of vertical extent or of cumulus type when the air rises rapidly, often because of instability. A simple classification based on base and type is given in Table 5.2.

Table 5.2. *A cloud classification.*

Group	*Base of cloud*	*Type*
High clouds indicated by prefix cirro	20,000 to 40,000 ft	Cirrus Cirro-cumulus Cirro-stratus
Middle clouds indicated by prefix alto	7,000 to 20,000 ft	Alto-cumulus Alto-stratus
Low clouds	Surface to 7,000 ft	Cumulus Cumulo-nimbus Strato-cumulus Stratus Nimbo-stratus

There are a large variety of cloud types which are described, named and classified in the International Cloud Atlas. Observers handbooks and cloud charts of the various meteorological services are useful for those who require further details.

REFERENCES

[1] London Met. Office, *Hygrometric Tables*, for the compilation of relative humidity, vapour pressure and dew point from readings of dry and wet bulb thermometers exposed in Stevenson Screens, H.M.S.O., London.

[2] United States Weather Bureau, *Psychrometric Tables*, WB No. 235 U.S. W.B., Washington, D.C.

[3] *Smithsonian Meteorological Tables* (1958). The Smithsonian Institution, Washington, D.C.

[4] KIEFER P.F., (1941). The thermodynamic properties of water vapour. *Monthly Weather Rev.* 69, (Nov.).

[5] SLATYER R.O. and McILROY I.C., (1961). *Practical Microclimatology*, U.N.E.S.C.O.

OTHER REFERENCES

BERRY E.A., BOLLAY E. and BEERS N.R., (1945). *Handbook of Meteorology*, McGraw Hill N.Y.

PENMAN H.L., (1955). *Humidity*. The Institute of Physics, London.

CHAPTER SIX

The meteorological elements — Winds

6.1 The occurrence of wind. Wind is air in motion. It occurs because of the uneven heating and cooling of the earth's surface. Wind transfers energy and water vapour throughout the atmosphere to adjust the disparities of temperature and water content on the globe. In the evaporation process, wind brings heat and water vapour into or out of a water surface. Wind is also an essential component part of a storm where it is related to the supply of moisture, the storm model, and the precipitation process itself. Other important effects of wind are the pressures exerted on structures or wind loadings, the vibration of objects under the influence of eddies or gusts, and the generation of waves and swell on a water surface. The relationships between the wind and the pressure or temperature fields given by isobaric and isothermal maps enable estimates to be made of wind in the absence of anemometers.

6.2 The measurement of wind. The horizontal component of the air movement, parallel to the earth's surface, is a measure of the wind in both direction and magnitude. The vertical components, often referred to as air currents, have resultants which are much smaller than the horizontal but have considerable significance in weather phenomena. The movement of air over a surface varying in properties such as temperature and roughness, produces eddies of many sizes and results in fluctuations of wind in both direction and velocity or wind gusts.

Wind direction is measured to eight or sixteen points of the compass or is expressed in degrees from north. The wind vane is the usual instrument and can be made to record the direction graphically. Also, the wind direction may be deduced from plumes of smoke or the movement of flags and trees. The direction given is the one from which the wind blows, that is, a north east wind is a wind blowing from the north east.

Wind velocities are measured by anemometers. The rotating propeller, or cup type, gives a value of the 'run of the wind' over a period. The pressure tube, or Dine's type, acts on the principle of the pitot tube and gives a value of the pressure exerted by the wind. This is converted to a reading of wind velocity over a few seconds and is often shown graphically. In this instrument, gusts and lulls are clearly shown. The inertia of the cup type instrument damps out these fluctuations.

The mean wind velocity over a period measures its power potential for wind mills or the rate of transport of heat and moisture for evaporation and storm studies. It is obtained from a cup anemometer by dividing the 'run of the wind' by the period over which this is taken. Mean velocity is estimated from the graphical record by averaging the curve through the gusts and lulls. Wind velocities may also be estimated by its effect on objects, the Beaufort wind scale being an example of this method.

The gustiness of a wind arises from the numerous eddies generated within the stream and is significant in storm damage and wind loadings on structures. A measure of gustiness is the range of wind velocities, that is, the difference between maximum and minimum velocities in a period. Gustiness is related to the stability of the air which depends upon the air's source region and the nature of the surface over which it has moved. Therefore, stability and gustiness vary with the wind direction. Unstable and gusty winds can be expected from some directions while stable and smooth flows come from other points of the compass. Stable air has a small range of gustiness while unstable air can show big variations.

Wind velocities are measured in kilometers per hour, metres per second, feet per second, miles per hour, or knots. Conversion factors are:

$$1 \text{ m/sec} = 1{\cdot}9 \text{ knots} = 2{\cdot}2 \text{ miles/hr} = 3{\cdot}2 \text{ ft/sec}$$

$$1 \text{ knot} = 1{\cdot}15 \text{ miles/hr and } 1 \text{ km/hr} = 0{\cdot}62 \text{ miles/hr.}$$

6.3 The relation of wind to pressure. Atmospheric pressure, air temperature, and density are inter-related through the gas law $p = RT\rho$. Moisture in the air is allowed for by using virtual temperature T_v. The gas law then becomes

$$p = RT_v\,\rho.$$

Changes of temperature in the atmosphere lead to changes of air density and so differences of pressure in the horizontal plane. Such differences give rise to air movement or wind. As air pressure can be conveniently measured with the greater accuracy, it is the practice to relate wind to the pressure pattern. This is best shown by an isobaric map, constructed by drawing isobars or lines joining places of equal mean sea level pressure. The inter-relationship with temperature and air density, however, should be borne in mind, and it is as reasonable to use the temperature pattern or the density or thickness pattern instead of that of pressure. Maps constructed to show upper air flow make use of these alternatives.

The forces acting on the atmosphere can be summarised as follows:

(*a*) A pressure force, B, proportional to the pressure gradient, and developed by temperature differences. Its direction is from high to low pressure and it is expressed simply as

$$B = \frac{1}{\rho}\frac{dp}{dx} \text{ (one dimensional case).} \qquad (6.1)$$

Wind flow on a non-rotating earth without friction could result from such a force.

(*b*) A force introduced by the earth's rotation with angular velocity, ω, or the geostrophic or Corioli's force, G, has already been discussed in the section on the general circulation. For the simple case this force may be expressed as

$$G = 2v.\ \omega\ \sin\phi. \qquad (6.2)$$

This shows its variation with latitude (ϕ) or its increase with distance from the equator.

(*c*) A force developed through air movement along a curved path or cyclostrophic or centripetal force, C, is proportional to v^2/r where r is the radius of curvature

i.e. $$C = \frac{v^2}{r} \tag{6.3}$$

(*d*) A frictional force dependent on the nature of the earth's surface. Above 2000 ft, or 600 m, this is small and the wind flow is determined by the other factors.

In the simple case of non-frictional, straight flow the last two forces may be neglected and the pressure force balances the Corioli's force. This is expressed in Buys Ballot's Law, and quantitatively,

$$2v\omega \sin\phi = \frac{1}{\rho}\frac{dp}{dx}$$

$$\text{or} \quad v = \frac{1}{\rho\ 2\omega\ \sin\phi}\frac{dp}{dx}\,. \tag{6.4}$$

Air flow is thus along the isobars and is proportional to the pressure gradient dp/dx or, if equal intervals are used for isobars, i.e. dp or $p_2 - p_1$ is constant, the velocity is inversely proportional to the distance apart of the isobars. This means simply that strong winds occur with a close spacing of the isobars. Also, it follows that latitude must be considered and the same pressure gradient will give higher velocities in the lower latitudes. These relationships are expressed in geostrophic wind scales which are often printed on the weather charts concerned as the projection and scale of these charts are relevant. Alternatively, tables may be used to find the geostrophic wind in terms of the variables involved. The geostrophic wind is thus the air movement resulting when the pressure and Corioli's forces are in balance. It applies to straight isobars or those of a slight curvature and to the freely flowing air above the earth's frictional layer.

The wind resulting from the balance between the pressure, Corioli's and cyclostrophic forces is called the gradient wind. It represents the wind flowing along curved isobars clear of the frictional influences of the earth's surface. This case is shown in Fig. 6.1 with the appropriate equations for the northern and southern hemispheres. In low latitudes the Corioli's force becomes small and the air movement is controlled by the pressure and cyclostrophic forces. The term cyclostrophic wind is given to the resulting high velocity current which is typical of a tropical cyclone.

By the use of vector analysis, the three dimensional case can be developed and suitably expressed in mathematical terms. Thus, the freely moving wind's speed and direction can be determined from isobaric maps and these relationships. They apply above the frictional layer of the earth's surface. Friction dissipates portion of the wind's energy and reduces the geostrophic force, so that the pressure gradient wind is somewhat greater and causes the wind in the surface layers to blow slightly out from a high and into a low.

Where friction is greatest, as over the land, this effect is more pronounced. It has been found that the surface wind over the sea is inclined, in middle latitudes, at approximately 15° to the low pressure system and is two thirds of the wind speed at 2000 ft. Over the land the angle of inclination is 25–30° and the surface speed approximately half that of the freely flowing air.

Estimates of surface winds, at 10 m, can thus be obtained from the gradient winds and

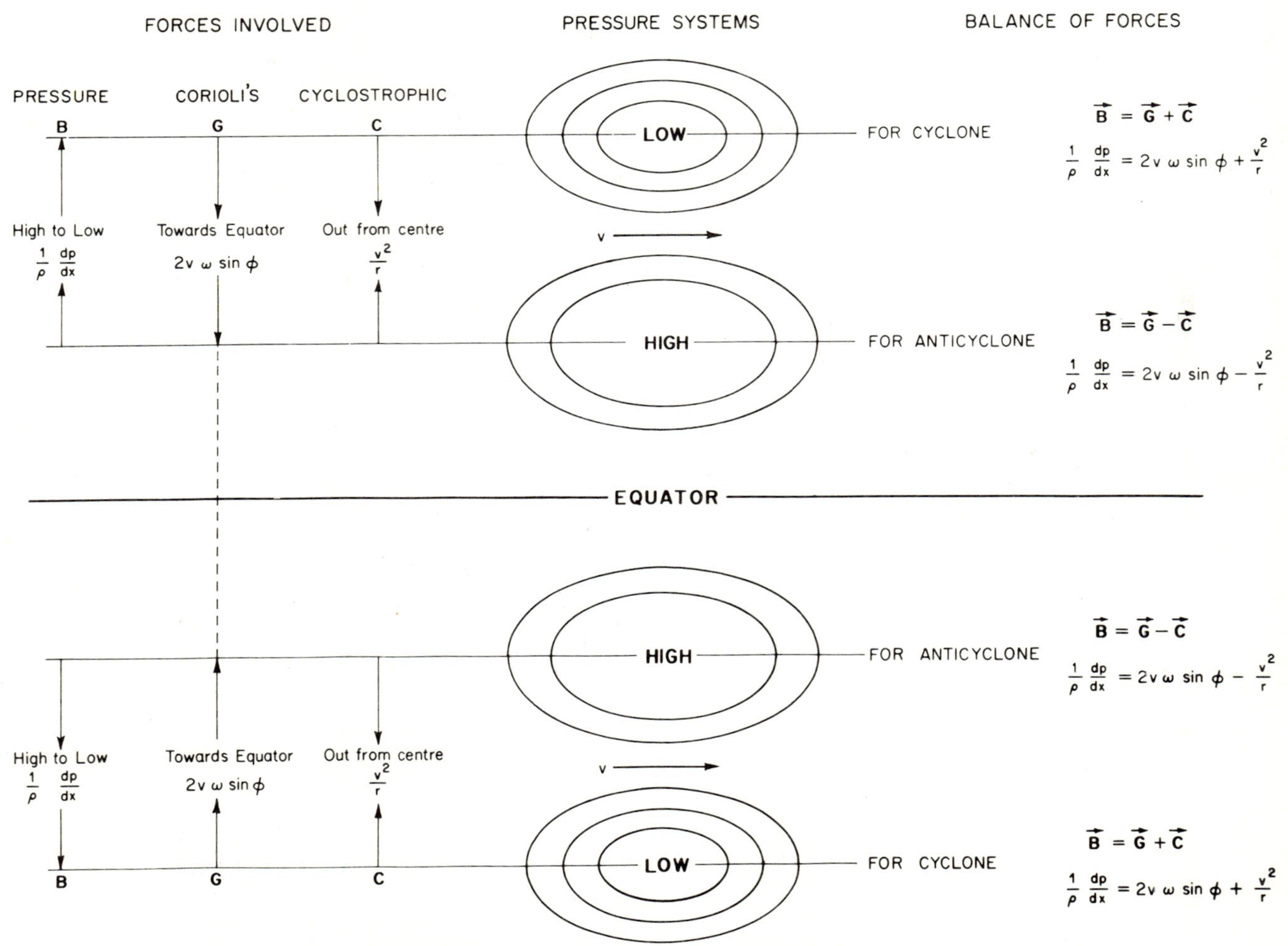

Fig. 6.1. Relationship between wind, pressure, and other variables.

so, from isobaric maps. For example, the mean monthly pressure maps give the average monthly gradient winds. By multiplying these by two thirds on the windward coasts, one half inland, and one third inside continents, a suitable estimate of the average monthly surface winds may be obtained.

6.4 The variation of wind with height. In the surface layers, the roughness and variations of the earth's surface and the stability of the air create physical and thermal barriers to the free movement of air and modify its trajectory. These effects cause the wind velocities in the boundary layer to vary in direction and magnitude with height above the earth's surface. The relationship found empirically between the gradient wind and the surface, or 10 m, wind has been mentioned in 6.3, but the wind profile near the ground will be further discussed.

Wind velocity is generally accepted as varying logarithmically with height,[1] and formulae of the form $v = a \log z + b$ and $v = a \log (z + c) + b$, where a, b and c are constants, can be established for sets of observations in particular sites where roughness is constant and when the atmosphere has a given stability. If the constants of these equations are evaluated from experimental data, winds at other heights may be estimated. Alternatively, the basic data may be plotted as velocity against the logarithm of height and a smooth line or curve drawn. Interpolation or extrapolation can then be made for the conditions of the original data which, however, must be observed at more than one height, and preferably, at many heights. Unfortunately, such data are not often available.

The variation of wind with height, can also be represented by a power profile of the form

$$\frac{V_1}{V_2} = \left(\frac{Z_1}{Z_2}\right)^k$$

or $v = cz^k$ where c and k are constants for set conditions. The value of k has been determined as 1/7 for a wide range of conditions, chiefly in the layer from 0–10 m but there are marked diurnal and seasonal variations. Over a snow surface for conditions assisting snow melt, a value of $k = 1/6$ has been proposed. As the air becomes more stable, the value of k tends to 1/2 with an average value of 1/5. The power profile is easier to use than the logarithmic relationship. Information [2] on the average wind speeds at two heights enables the determination of k or evaluation of $v = cz^k$. For example, if the wind speed at 16 m is 4·69 m/sec and at 32 m is 5·40 m/sec the equation becomes $v = 2{\cdot}68z^{0{\cdot}202}$. Here the value of k approaches the average value 1/5 suggested above.

Wind gust velocities also vary with height and average ratios obtained experimentally [3] show, in Table 6.1, the relationship between gusts at heights of 10, 60, 100 and 300 ft to gusts at 33 ft for different sites.

Table 6.1. *Average ratios of wind gust velocities at stated heights to gust velocities at 33 ft.*

	Height (ft)				
Location	10	33	60	100	300
On coast	0·93	1·00	1·06	1·09	1·22
Inland	0·97	1·00	1·03	1·05	1·14

6.5 Influence of physiography on wind. The earth's surface affects the wind's flow other than in the dispersal of energy by friction. The flow patterns or stream lines of the overall circulation or weather system may be distorted by obstacles in their path. Also, the varied nature of the earth's surface introduces thermal effects which modify the gradient winds.

Hilly and mountainous country are responsible for many convergence and divergence effects. The wind speeds above hills are considerably greater than at the same height without the obstruction due to this converging action. Similarly, the winds are correspondingly reduced on the lee side although eddies give erratic readings under some situations. The highest recorded wind velocities have been reported from the tops of mountains where wind loadings are correspondingly greater than in valleys. Wind generators are also best placed on hills of an aerofoil profile to take advantage of increased wind velocities without undue turbulence caused by sudden rises and falls in the land surface; also, tunnel shaped valleys cause convergence of the horizontal flow with increased velocities. Combined with the usual rise at the head of a valley, this can give increased wind velocities and vertical lift which, with suitable air masses, result in heavy rain.

The height to which the normal wind flow is disturbed by topography depends upon many factors including the roughness and nature of the surface, the stability of the air, the direction and speed of the wind, as well as the height and slope of the deflecting surface. If wind measurements with height are available above the mountains and the adjacent windward plain, the converging effect may be deduced by comparing winds at the same heights in both locations. The elevation where wind velocities are similar marks the limit of the orographical influence or the nodal surface. Alternatively, if the height of the nodal surface is known or estimated, streamlines of wind flow may be constructed, and wind velocities at different heights over the mountain can be computed using the principle of continuity. A rough estimate of the height of the nodal surface is $2\frac{1}{2}$ times the height of the barrier.

The vertical wind component introduced by a barrier is $w = v \sin \alpha$ where α is the angle of slope, and v and w the horizontal and vertical velocities. This applies near to the surface, and drops off exponentially with height. A factor of $\frac{1}{2}$ every 1000 m has been suggested as the rate of decrease of orographical lift with height. [4] The barrier is assumed at right angles to the horizontal wind. If it is not, the component at right angles is taken.

6.6 Thermal effects and local winds. Differential heating of the earth's surface occurs because of its varied nature. On occasions, thermal gradients are strong enough to initiate wind movements or modify the prevailing circulation and are discussed briefly in chapter 3. The commonest local winds are the land and sea breezes encountered in coastal regions. Under insolation during the day, land reaches a higher temperature than the adjacent sea. Air above the land becomes heated with density and pressure differences developing between this air and that over the sea. When a sufficient thermal density or pressure difference has developed, air moves in from the sea and continues as long as gradients are maintained. The velocity of the sea breeze depends upon temperature contrast and so upon latitude or insolation, as well as on the nature and physiography of the coast. The sea breeze can extend inland for 40 miles or more and to a height of 1000 ft. The velocities can exceed 20 knots and are often strong enough to overcome or modify a gradient wind. Due to the marked insolation and small Corioli's force in lower latitudes, sea breezes occur regularly in the tropics, and by lowering temperatures have a great influence on physical comfort.

During the night or early morning, the thermal gradients decrease or become reversed with the dying out of the sea breeze, or the initiation of an air movement from land to sea, called the land breeze. As the temperature contrasts are usually less marked, the land breeze is lighter and does not extend far beyond the immediate coastline. In some localities, when reinforced by topographical or gradient winds, the land breeze can add to or start off a strong offshore wind in the early morning.

Temperature differences also develop in mountainous and hilly regions and initiate local winds. In the morning, the more rapid heating of favoured upper slopes may cause air in contact with them to become warmer than that upon nearby lower areas, and results in a gentle flow of air up the slope as an anabatic wind. It ceases when the temperature contrast is lost. In the evening, the cooling of air on the higher slopes at a quicker rate than in the nearby valleys, can give a thermal gradient which results in a flow of cooler air down the slope as a katabatic, gully, or valley wind. Winds of this type are common and are given local names. Their effects are usually confined to the surface layers and influence physical comfort rather than produce rainfall.

6.7 Change in wind's properties. An air mass passing over a surface is modified in many ways, depending on the nature of the surface and the time the air is over it. Temperature, humidity, and stability may all be affected. These changes are discussed under air masses.

Also, wind passing over a mountain range is modified by the so called 'Föhn' effect. When air under the influence of a pressure system is forced to ascend a mountain range, it cools at the D.A.L.R. until saturated and then at the S.A.L.R. cloud and precipitation often form, with the removal of some moisture from the air mass. As the air descends on the lee side it increases in temperature at the D.A.L.R. and as cloud is not formed on descent, greater insolation is received by the land surface. The air becomes warm and dry and is known as a Föhn wind. It can damage crops and give unpleasant living conditions while it continues to blow.

6.8 Shelter belts. The frictional effect of objects on the earth's surface is used to reduce wind velocities and modify climate. Belts of trees at right angles to the prevailing wind and forests are successful in damping out wind in the surface layers. Within a forest, the wind quickly decreases in velocity at a rate depending on the density of cover. This leads to a modification of the climate within the forest. Trees give protection on their lee side to a distance of about 20 times their height with a measurable reduction within a distance of 50 times their height.

On the windward side there is protection for about 4 times the height of the trees. The wind flow over a region can be modified in this way by the appropriate siting of forests and belts of trees normal to the persistent winds. Modification of the radiation balance and the interception and retention of water from fog, mist, cloud and storms are also important hydrological effects of trees.

Brush fences, mounds, and furrows also have important influences on the wind flow in the boundary layer and accordingly affect microclimate and the response of living organisms in these areas. [5]

REFERENCES

[1] BRUNT D., (1944). *Physical and Dynamical Meteorology,* Cambridge.

[2] BROOKS C.E.P. and CARRUTHERS N., (1953). *Handbook of Statistical Methods in Meteorology,* H.M.S.O., London.

[3] CARRUTHERS N., (1943). Variation in wind velocity near the ground. *Q. Jl. R. Met. Soc.* **69**, 289.

[4] KNOX J.B., (1960). Procedures for estimating maximum possible precipitation. *Calif. State Dept., Water Resources Bull.* 88.

[5] GEIGER R., (1965). *The Climate Near the Ground,* Harvard University Press, Cambridge, Mass.

CHAPTER SEVEN

The meteorological elements — The pressure of the atmosphere

7.1 Behaviour of the atmosphere. In describing the physical state of the atmosphere, the elements of temperature, humidity and wind, together with their relationships with pressure, have been discussed and expressed in the gas laws and in the gradient wind equations of Fig. 6.1. These draw attention to the importance of pressure measurements and its variations with height, time and area. The study of these leads to the extensive fields of altimetry, barometry and synoptic meteorology. Here, only a few aspects of these subjects will be considered.

7.2 Measurement of pressure. Atmosphere pressure is commonly measured by a mercurial barometer, in which the total weight of the column of air above the barometer effects a pressure which is balanced by the weight of the column of mercury. The height of this column is thus a convenient measure of air pressure and units of inches or millimetres of mercury are used. The basic pressure unit however is the bar which is defined as a pressure of 10^6 dyn/cm^2. A millibar (mb) is 1/1000 bar or 10^3 dyn/cm^2. Conversion factors are 1 bar = 1000 mb = 29·531 in of Hg = 750·1 mm of Hg.

Variations in pressure, as distinct from absolute values, are shown by an aneroid barometer or barograph. These instruments contain an aneroid cell which is an airtight metal box almost exhausted of air. Its movement in and out as the atmospheric pressure rises and falls moves a hand over a dial in the aneroid barometer, or a pen across a paper marked with a time scale and on a rotating drum, in the barograph.

7.3 Variation of atmospheric pressure with height. Atmospheric pressure falls off with height as there is less air above the instrument. An approximate figure is 1 mb for 30 ft or 9 m near the earth's surface, but this depends upon temperature and pressure as follows.

If at height z, and pressure, p, temperature, T, and density, ρ, in a unit column of air, a disc of air, dz, is taken then $p + dp$ is the pressure at height $z + dz$

and
$$dp = -g\rho\, dz \tag{7.1}$$
or
$$\frac{dp}{dz} = -g\rho = \frac{-gp}{RT}\,. \tag{7.2}$$

This altimetry equation is the basic relationship between height and the atmospheric variables. The variation of gravity (g) within the heights normally used in meteorology is insufficient to be taken into account and may be assumed constant.

If the variation of density or temperature with height is known in terms of p and z then integration gives a relationship between pressure and height and enables pressure to be used as a measure of height as in the altimeter.

Assuming an isothermal atmosphere (T constant), equation 7.2 becomes

$$\frac{1}{p}\frac{dp}{dz} = \frac{-g}{RT} \text{ where } \frac{-g}{RT} \text{ is constant.}$$

Integrating $\log p_0 - \log p = \frac{gz}{RT}$ p_0 is pressure at $z = 0$ (7.3)

$$z = \frac{RT}{g}\log\frac{p_0}{p} \quad (R = 2{\cdot}87 \times 10^6) \tag{7.4}$$

$$z = 67{\cdot}4\, T \log_{10}\frac{p_0}{p} \quad (z \text{ in m}). \tag{7.5}$$

Assuming a standard temperature T of 10°C a scale of an altimeter can be constructed in accordance with equation 7.4 or 7.5 and gives readings of height as the instrument is moved above the base level.

If the mean temperature of the layer differs from 10°C a correction to the height may be made by increasing the height estimated by the altimeter by 1/283 for every degree rise in temperature of the layer. Humidity can also be allowed for by using the virtual temperature.

Assuming an atmosphere with constant lapse rate (β), temperature at height z, $T = T_0 - \beta z$ where T_0 is temperature at p_0, equation 7.2 becomes

$$\frac{1}{p}\frac{dp}{dz} = \frac{-g}{R\,(T_0 - \beta z)}\,. \tag{7.6}$$

Integrating with pressure p_0 at $z = 0$

$$\log\frac{p}{p_0} = \frac{g}{R\beta}\log\frac{T_0 - \beta z}{T_0} \tag{7.7}$$

or

$$\frac{T_0 - \beta z}{T_0} = \left(\frac{p}{p_0}\right)^{\frac{R\beta}{g}} = \left(\frac{p}{p_0}\right)^{0{\cdot}293\beta}$$

where β is in °C/100 m. (7.8)

This relates height to pressure for any desired surface temperature T (°K) and lapse rate B (°C/100 m).

The International Commission for Air Navigation (I.C.A.N.) specified an atmosphere up to 11 km as $p_0 = 1013{\cdot}2$ mb, $T_0 = 288$°K and $\beta = 0{\cdot}65$°C/100 m.

Equation 7.8 becomes

$$p = 1013{\cdot}2\left(\frac{288 - 0{\cdot}0065z}{288}\right)^{5{\cdot}256} \quad \text{for } z \text{ in m.} \tag{7.9}$$

Table 7.1 shows some values for the I.C.A.N. atmosphere.

The I.C.A.N. specified that above 11 km an isothermal atmosphere with constant temperature of −56·5°C be used. Equation 7.5 is applicable. Altimeters are constructed with scales in accordance with the I.C.A.N. specifications. Corrections may be made for conditions differing from the assumptions but, in practice, are not large.

Table 7.1. *Relationship between height, temperature, pressure and density for the I.C.A.N. atmosphere.*

Height m	Temperature °C	Pressure mb	Density kg/cbm
0	15·0	1013·2	1·225
1000	8·5	898·7	1·112
2000	2·0	794·9	1·007
4000	−11·0	616·3	0·819
8000	−37·0	355·8	0·525
12000	−56·5	192·2	0·309

7.4 Variations in atmospheric pressure with time. Diurnal variations occur in air pressure with maxima about 10 a.m. and 10 p.m., and minima about 4 a.m. and 4 p.m. These are associated with a double wave in the atmosphere passing around the earth about two hours ahead of the sun. The magnitude of the diurnal oscillation is as much as 4 mb at the equator, 1 mb at latitude 50° and nil at the poles. In higher latitudes, the smaller diurnal variations are masked by the passage of high and low pressure systems, but, in the tropics the larger range and lower frequency of pressure systems show the pressure wave distinctly.

The relationship between pressure, temperature, and density has already been considered. Under heating the sequence is probably a rise of temperature, reduction of density over the heated regions, and air movement from the colder areas where air pressure is higher. Thus high pressures are associated with the lower temperatures and vice versa. Annual changes in atmospheric pressure are thus the reverse of the temperature regime. In winter time when the continents are relatively cold, high pressures develop with lower pressures over the warmer oceans. In summer, high pressure persists over the comparatively cooler sea areas with low pressures over the continents. Mean monthly pressure maps show these features clearly.

The general circulation, already described, gives a possible distribution of winds and pressures on a homogeneous rotating earth to restore the thermal balance. The temperature effect on pressure which occurs because of the non-homogeneity of the earth's surface, described above, explains the seasonal shift of the pressure systems between oceans and continents.

7.5 Pressure fields or isobaric maps. Compared with other meteorological variables, air pressure can be measured with considerable accuracy, to at least five significant figures. Accordingly, the state of the atmosphere can be comparatively well defined by its pressure distribution. As pressure readings are taken at different heights they must be corrected to a standard level, usually mean sea level, (M.S.L.). This is done by assuming a temperature for the column of air from M.S.L. to the station and applying a correction. Mean sea level pressures made at the one time, or synoptic reports, are plotted on maps and places of equal pressure joined by isobars. Patterns of high and low pressure are shown and are often associated with certain types of weather and can have characteristic movements. The subject of synoptic meteorology studies these systems and has evolved methods of weather forecasting. A few of the simpler types of pressure patterns are described below.

7.6 High pressure areas or anticyclones are regions of relatively high pressure. They are generally elliptical in shape but may possess *ridges* extending into preferred areas. They can span twenty degrees or more of latitude and longitude but are sometimes quite small. The wind circulation is clockwise in the northern hemisphere and anti-clockwise in the southern.

The general circulation, already described, give reasons for high pressure areas about latitude 30° and the polar regions on a homogeneous planet. This is disturbed by lack of homogeneity but the broad pattern may be seen in mean pressure maps and most clearly in the southern hemisphere where conditions are more uniform. The movement of anticyclones along tracks near latitude 30° in winter and latitude 40° in summer have important influences on the weather of these areas. Near the centres of high pressures, subsidence or descending air is associated with fine weather. However, at the edges of the anticyclones, comparatively strong streams of moist air may be directed onto the coasts or mountain ranges to produce cloudy and showery conditions.

7.7 Tropical cyclones. These are regions of low pressure occurring within the tropics. They may also be called hurricanes, typhoons, or by local names. They are extremely intense depressions with M.S.L. pressures sometimes below 915 mbs (or 27 in) and diameters from one to five degrees of latitude. During the summer they originate in the open ocean between latitudes 5–15° and move on irregular paths, usually towards the higher latitudes. They are associated with gale force winds, torrential rain, and cause great damage. They increase in size as they develop but lose intensity when they move over land away from their source of energy, namely the water vapour evaporated from the ocean. The isobars are closely spaced and winds are anti-clockwise in the northern but clockwise in the southern hemisphere.

7.8 Extratropical cyclones or depressions are also regions of low pressure which form and develop outside the tropics or in the middle latitudes. This is the polar front region where air masses of contrasting properties from the poles and equatorial regions are brought together with considerable exchange of energy. As with the anticyclones they have a general movement from west to east. Their mean paths move seasonally northwards or southwards with the thermal equator. They possess a strong wind circulation which is anticlockwise in the northern hemisphere but clockwise in the southern. They are often associated with extensive precipitation and cloud and may cover anything from a one to thirty degrees. Although the rate of precipitation may be less than in the tropical cyclone, rain may continue for long periods of time and cover large areas. Strong winds, high seas, and heavy swells are common features of these high latitude storms.

7.9 Thermal low pressure areas appear on weather maps over continental areas in summer, especially during afternoons when the diurnal pressure wave is at a minimum and intense heating can give a fallacious pressure correction to mean sea level. These systems do not have significant wind circulations or weather phenomena.

7.10 Troughs of low pressure are distortions of isobars in a stream into a V or Λ shape. They may be extensions of a larger low pressure area or perturbations within a stream. In the prevailing westerly stream of middle latitudes there may be a number of these troughs or perturbations moving in the direction of the stream. The easterly flow of the tropics can also show a series of these tropical troughs, chiefly in the summer season. Along the axes of these troughs air may be introduced from the tropics or mid latitudes and converge with the existing air or air from other sources which is moving under inertia. In this convergence area, vertical motion occurs with thunderstorms, precipitation, heavy at times, and local squalls. Small lows or rain depressions can develop at the tip of the trough. The waves or troughs in the westerly are also marked by precipitation, cloud, and change of direction of wind, but have not the intensity or significance of the tropical troughs.

7.11 Cols are areas of intermediate pressure between high or low pressure systems. Winds within them are usually indefinite with air masses drifting in from widely differing sources, say, from the adjacent anticyclones. Weather developments occur within cols with the birth of cyclones or, cyclogenesis and the generation of frontal systems or, frontogenesis.

7.12 Pressure as an indication of weather. In latitudes where there are numerous depressions, about 40° in winter and 60° in summer, a fall in barometric pressure generally indicates the approach of an extra tropical cyclone or depression with strong winds and rainy weather. Rising pressure indicates the coming of an anticyclone with fine weather, clear skies and lighter winds. In these latitudes therefore, the variations of barometric pressure can be a useful indication of future weather. Some barometers are therefore marked directly in terms of weather as well as pressure.

In other latitudes, the relationship is not so clear. In the tropics the approach of a tropical cyclone is preceded by rapid falls of air pressure and the barometer serves a valuable service. However, quite frequently, rising pressure ahead of an anticyclone may mean increasing pressure gradients and wind with a strong stream of moist trade wind air which may give frequent showers. Pressure tendencies must therefore be considered with caution when used for forecasting weather.

REFERENCES

[1] International Commission for Air Navigation (Dec. 1938). *Official Bull. No. 24,* Res. No. 1053.

OTHER REFERENCES

PETTERSSEN S., (1958). *Introduction to Meteorology,* McGraw Hill, N.Y.

BYERS H.R., (1959). *General Meteorology,* McGraw Hill, N.Y.

RIEHL H., (1954). *Tropical Meteorology,* McGraw Hill, N.Y.

CHAPTER EIGHT

Air masses and fronts

8.1 Air mass sources. The triple cell circulation suggests a probable movement of air over the earth's surface and one which satisfies the known thermal and dynamical considerations. This circulation is verified by the mean maps of the wind and pressure distribution over the globe. The presence and nature of land and sea modify these principal air movements and introduce many secondary circulations which are shown in daily weather maps and by local and diurnal winds.

Stagnant periods of light and variable winds occur over parts of the earth's surface. These are associated with high pressure areas which persist over favoured localities which are usually continents in the winter and warm seas in the summer. Thus, large bodies of air remain, for some time, in contact with portions of the earth's surface and experience comparatively uniform meteorological conditions to become relatively homogeneous, at least in horizontal extent. In this way are generated air masses with surface temperatures and humidities approaching these of the underlying surface. Visibility, cloud, precipitation forms, and lapse rates of temperature and humidity, which are also associated with turbulence and the gustiness of winds, become similar throughout the air mass.

The localities where air masses originate are known as source regions. They may be land or sea areas to give rise respectively to continental (c) or maritime (m) air masses. The continental air mass is usually dry and fluctuates widely in temperature from relatively cold in winter to hot in summer. The maritime air mass is nearly always moist with a more uniform temperature. Its dew point approximates to the sea temperature where it originates.

The source regions may be within the Equatorial (E), Tropical (T), Polar (P), or Arctic (A) latitudes. This gives a further basis of classification. In particular, the subtropical high pressure cells of the general circulation are source regions for tropical air and the Arctic Circle highs for the arctic air. Polar air originates in the higher latitudes. Equatorial air may be distinguished from tropical air in that it acquires high temperatures and moisture in the doldrums and usually originates from the opposite hemisphere. It is the air mass involved in the summer monsoon.

8.2. Classification of air masses. A simple classification of air masses is given in the following table.

Air masses are given local names after the oceans or land masses where they originate. Air masses are identified by their properties and the characteristics of their places of origin. The weather conditions likely to occur can be deduced from these properties and the passage of the air mass from its source into other areas. Weather processes which

Type of air mass	Symbol	Source	Properties
Equatorial	*E*	Doldrums	Very warm, moist, unstable
Tropical maritime	*Tm*	Ocean; sub-tropical high	Warm, moist, varied stability
Tropical continental	*Tc*	Land; sub-tropical high	Dry, hot in summer, cold in winter. Often stable below 2 km but unstable above
Polar maritime	*Pm*	Ocean; mid latitudes	Cool, moist, rel. unstable
Polar continental	*Pc*	Land; mid latitudes	Cool, dry, unstable
Arctic maritime	*Am*	Ocean; Arctic high	Very cold, moist, stable
Arctic continental	*Ac*	Land; Arctic high	Very cold, dry, stable

occur within an air mass are turbulence, convection, and convergence. When the air mass meets a rising land surface topographical uplift occurs and, when it meets another air mass, frontal processes are expected.

8.3 Modification of air masses. On leaving their source regions, air masses become modified. The extent of modification depends upon:

(*a*) The nature of the surface over which they move. If this surface is warm, the air is heated from below and tends to instability. If the surface is cool, the air is cooled from below and becomes more stable. If the surface is moist, water vapour is added by evaporation to an extent dependent on the vapour pressure gradient. If the surface is dry, water vapour may be condensed from the air masses which is thereby dessicated. If steep hills are passed over, precipitation processes may be initiated and water removed from the air mass. Descending motion may heat the air mass and enhance evaporation.

(*b*) The time spent by the air mass passing over the surface.

(*c*) The processes which occur within the air mass. Turbulence and convection may occur because of increased instability, and lead to cloud formation and precipitation with removal of water from the air mass. Turbulence may also cause the downward mixing of dry air aloft and increased evaporation from the surface with an increase of moisture in the air mass and greater uniformity. The accumulation of air called convergence or its converse, divergence, in the surface layers may have similar effects causing uplift or sinking and so, removing water from the air mass in precipitation or, taking up water by evaporation.

(*d*) Conditions aloft affecting the air mass. For example, if an air mass cools from above, increased instability results or, if the air mass is warmed from above, it becomes more stable. It follows that many of the factors and processes which modify air masses are inter-related. As a general rule, air masses which move from the poles to the equator pass over warmer areas, to be warmed from below, and become less stable. Also, air masses moving into higher latitudes, pass over colder areas to be cooled from below, and become more stable. As exceptions, land in winter is often colder than neighbouring seas and has a stabilising influence while land in summer is sometimes warmer than the maritime air passing over it and in this case increases its instability.

8.4 Frontal surfaces and fronts. Air masses of widely differing properties exist together on the earth's surface and move under forces initiated by uneven heating. The boundary between dissimilar air masses is known as a frontal surface. When flow in the dissimilar air masses is parallel or divergent the surface is broad and diffuse but, when winds converge, air masses from differing source regions are brought into juxta-position and a distinct boundary or front is formed. Across the frontal surfaces or fronts occur changes in temperature, water content, wind, and pressure. The degree of discontinuity, density differences, and wind shear is related to the slope of the front.

Over the surface of the globe, the most clearly marked discontinuities occur in the higher latitudes between the polar high and subtropical high pressure regions. Here, Tropical, Polar, and Arctic air masses may meet and converge along a frontal surface known as the Polar front. On this surface of discontinuity, active fronts form. Low pressure waves are associated with the fronts and develop into extratropical cyclones with distinctive movements and life histories.

Frontogenesis, or the creation of distinct warm and cold fronts upon a stationary frontal surface thus occurs when two air masses converge, or the different density wind streams no longer run parallel to but bulge into each other. The convergence below and divergence aloft results in a drop of pressure with a low, cyclone, or wave developing. This process is called cyclogenesis.

8.5 The history of a polar front cyclone. The sequence of changes which take place on the polar frontal surface with frontogenesis and cyclogenesis is shown in Fig. 8.1.

The various stages are:

Stage 1. Warm and cold air flow in parallel streams separated by a frontal surface or stationary front.

Stage 2. Convergence occurs and flow is no longer parallel. The cold air bulges into the warm air with formation of a cold front. The warm air glides over the cold forming a warm front. At the intersection of the warm and cold front there is a fall of pressure and formation of a cyclone.

Stage 3. The cyclonic wave deepens with pressure falls at the vertex and the circulation becomes more vigorous. The lighter warm air continues to slide over the cold wedge of air and is lifted to form warm frontal cloud and rain. The denser cold air undercuts the warm air as a wedge and forces it to rise with possible cloud and showers.

Stage 4. The air associated with the front of greater velocity overtakes the other and

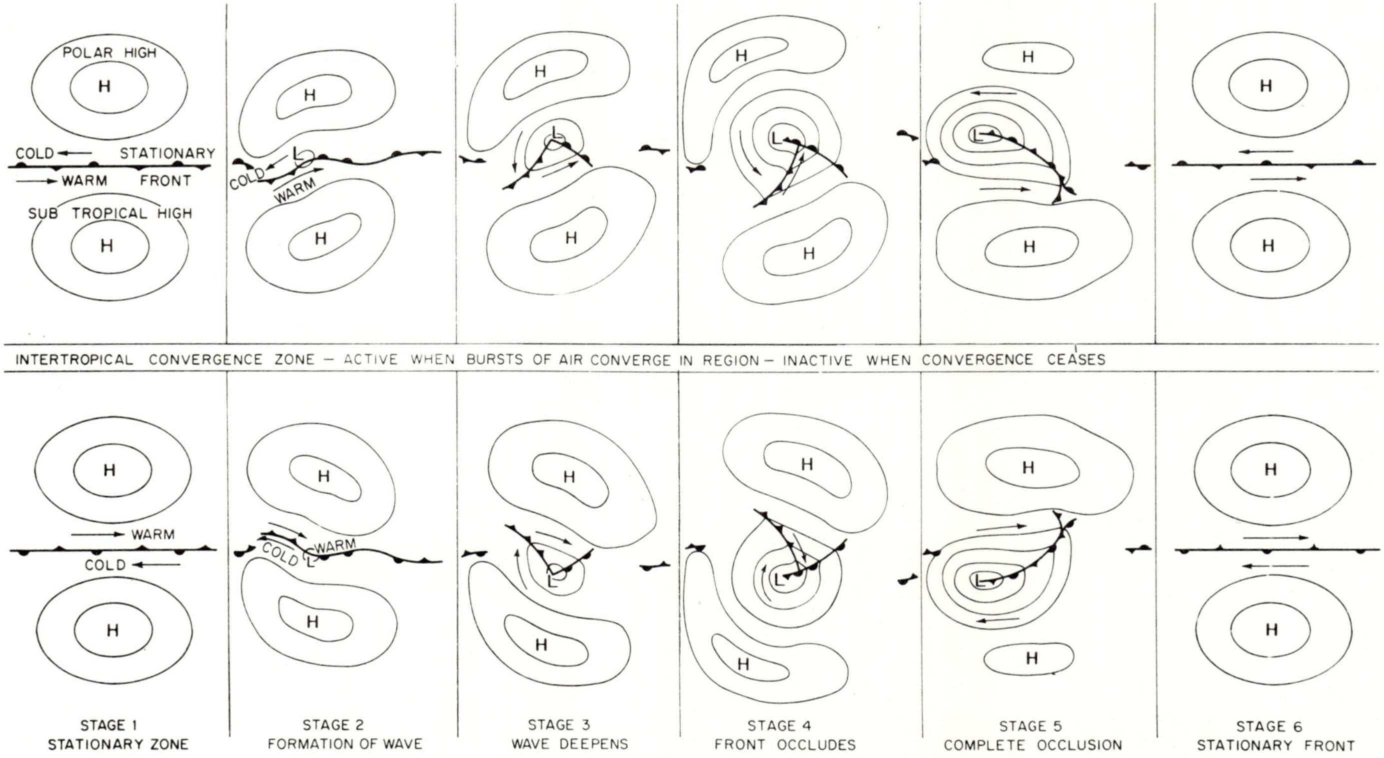

Fig. 8.1. The history of a polar front cyclone.

lifts it to give an upper pool of air. This is an occlusion. In the case given the warm front is occluded or lifted.

Stage 5. The occlusion reaches a more advanced stage with deepening pressures around the occluded portion and a stronger and more varied air flow in the zone of greatest convergence.

Stage 6. The air masses have become mixed, with less contrast and loss of wind shear, the frontal surfaces are destroyed. This is termed frontolysis. Also, the cyclone fills with the situation approaching stage 1. If conditions are suitable, further fronts and waves may be formed and follow a similar history.

8.6 Clouds and weather associated with fronts. The clouds and weather which develop at fronts depend upon the properties of the air masses involved and their relative velocities. The more contrast there is in properties between adjacent air masses, the steeper the frontal slope and the greater the convergence with correspondingly deeper cloud and heavier precipitation. The stability and water content of these air masses are also significant. A stable, dry air mass lifted by a front often gives little or no cloud and rain whereas an unstable, moist air mass can give deep cloud and heavy rain when forced to rise. Frontal action often 'triggers off' latene instability, that is, an initially stable air mass may become unstable when lifted by another air mass.

Modification of air masses by travel over warmer, colder, dryer, or moister regions become important in frontal action. For example, cold air, originating in higher latitudes, becomes unstable when moved towards the warmer areas near the equator. Similarly, warm air becomes more stable as it moves away from the equator into colder locations. Thus both origin and trajectory need to be considered to assess the properties of air masses involved in frontal action and the consequent weather. The Polar front region of the middle latitudes necessitates travel of the air masses from the sources regions to its north and south.

At a warm frontal surface it is usual for the relatively stable and moist, warm air to override smoothly the retreating cooler air mass. Slope of the surface is from 1 in 100 to 1 in 350. The vertical lift produced gives sheet cloud which appears as fine cirrus, thickens to cirrostratus and then, to altostratus and nimbostratus, from which rain may fall over a broad area ahead of the warm front. If the warm air is unstable, lifting may release latent instability with deeper cumuliform clouds imbedded among the sheet type to give additional localised heavy falls.

The cold front, with stronger wind velocities and greater density contrast, has a slope of about 1 in 50 and wedges into the warm air which it forces upward. The rapid lift can cause deep cumuliform cloud and heavy showers or thunderstorms to develop, especially when the warm air is moist and unstable. Intense rain may fall over a comparatively narrow band along the front where the lift is occurring. Ground friction retards the surface wind of the front giving an over-running nose to the cold front. Figure 8.2 shows the plan and a typical cross section of a Polar front cyclone.

Stationary and occluded fronts give a wide range of weather conditions which vary from light drizzle to heavy rain depending on the nature of the air masses, the vertical lift induced by their relative velocities, and the maturity of the system. The surface front sometimes remains stationary while the upper air flow continues to give prolonged cloud

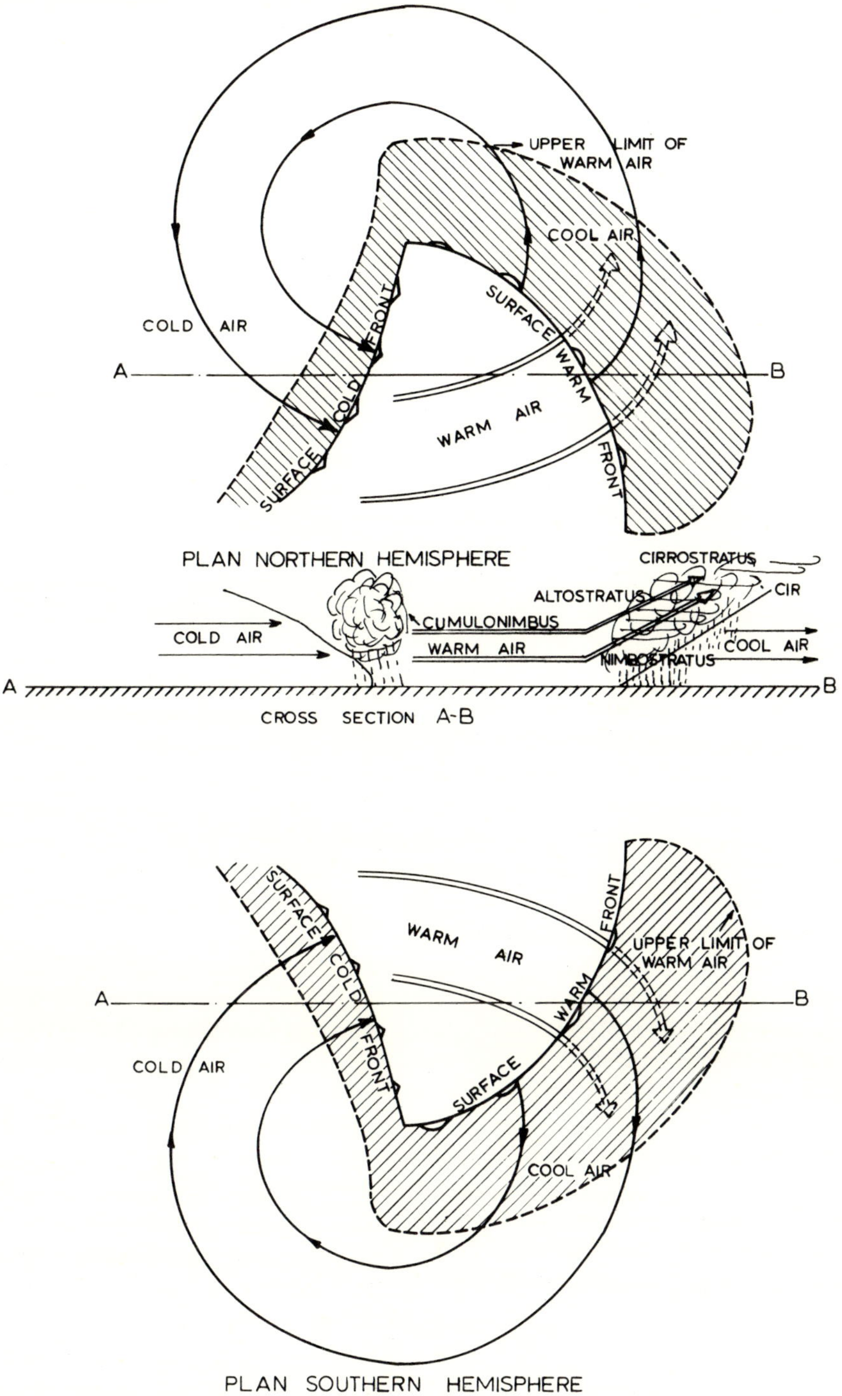

Fig. 8.2. Plan and cross section of a polar front cyclone.

and rain.

On the polar front, a series of extratropical frontal cyclones can form and go through their life cycles. They vary in size from minor waves of 100 square miles and lasting a few hours, to extensive depressions covering ten of thousands of square miles and persisting for many days.

8.7 Converging air masses in other areas. As well as at the polar front, the various air masses converge in other latitudes with appropriate discontinuities of the meteorological elements and weather phenomena. Also, local fronts form when favourable physiography encourages the development of a localised air mass which subsequently moves into an area where air of different properties is present. So called sea and land breeze fronts are examples of this action and involve cooler air undercutting air of higher temperature and different water content and stability.

Within the tropics, air masses meet in the intertropical convergence zone. Into this area of lower pressure, or the doldrums, tropical air masses from the subtropical high pressure cells, the NE and SE trade winds, and equatorial air, or the SW and NW monsoons converge.

Tropical air masses do not show much contrast in surface properties. Differences are mainly in their vertical structure, particularly moisture content and stability. Air masses in the tropics often move under local temperature effects or under inertia acquired in higher latitudes. When these effects are strong because of marked heating or intense subtropical anticyclones, large volumes of air move into the doldrums area and convergence occurs. Deep cloud and heavy precipitation then develops in this intertropical convergence zone. When wind velocities decrease due to lower temperature gradients or, to the weakening of the high pressure cells which drive the air into the equatorial belt, convergence decreases and the intertropical convergence zones disappear or show little activity and weather phenomena. As the air masses involved are frequently unstable and of low contrast, frontal surfaces in the tropics are not as distinct as in the polar front. The term convergence zone is to be preferred.

The seasonal march of the thermal equator produces a similar north–south movement of the triple cell circulation and the doldrums region. Thus the intertropical convergence zone has a seasonal movement with its furthest departure from the equator in the respective northern and southern summers. When the zone is more than 8 degrees from the equator, some Corioli's force is present and, with a favourable combination or convergence of air masses in this region, that is, the meeting of equatorial, tropical and modified polar air masses, tropical cyclones may form and develop. Such cyclones, because of the unstable nature of the air involved, do not have the distinct frontal structure of the cyclones of higher latitudes. Bands or belts of deep convective cloud surround the eye of the storm which is clear and calm. Outside this centre, strong gusty winds and heavy rains can cause immense damage and serious flooding.

REFERENCES

PETTERSSEN S., (1956). *Weather Analysis and Forecasting,* McGraw Hill, N.Y., Vols. 1 and Vol. 11.

RIEHL H., (1954). *Tropical Meteorology,* McGraw Hill, N.Y.

CHAPTER NINE

Evaporation

9.1 Evaporation and the hydrological cycle. The conversion of water from the liquid or solid state into the gaseous phase and its diffusion into the atmosphere is termed evaporation. As an essential part of the hydrological cycle, it redistributes heat energy between surfaces and the atmosphere.

Evaporation involves a supply of energy for the latent heats of fusion and vaporisation and a transfer process, which may be partly molecular but is dominantly turbulent.

Evaporation commences and continues, but not necessarily at the same rate, as long as there is a supply of moisture, a vapour pressure gradient between the water surface and the atmosphere and a supply of energy. The energy or latent heat is withdrawn from the most convenient source which may be the sun, the overlying air, the ground, or from the water itself.

In evaporation, solar radiation is the only independent factor but, with the other factors, reaches a balanced state with an appropriate rate of water vapour diffusion. A change in one factor is followed by adjustments in the others, perhaps, to reach a new balance or rate of evaporation. For example, if the sun is covered by a cloud, there is an immediate drop in the radiation upon a surface, but the evaporation rate changes more slowly because the vapour pressure gradient existing permits the continued diffusion of water vapour into the air. Heat for this is extracted from sources other than the solar beam, such as the surface itself or the air. This gives a drop in temperature of air and surface with a readjustment of the vapour pressure gradients to give lower evaporation and latent heat demand.

The reverse process to evaporation, or the conversion of vapour back into liquid or solid in the atmosphere, leads to the formation of cloud and fog. This, and a further phase, is initiated when the moisture-charged air is operated on by a storm mechanism which causes continuing condensation and results in the water reaching the ground as precipitation. This process is a spasmodic one, in contrast to evaporation which proceeds more or less continuously.

Water vapour diffuses from many different surfaces and, although distinctions are often made between water losses from water surfaces, damp soil, and plants, there is no fundamental difference in the physics of the process, but only in the nature of the surfaces. The physics of evaporation therefore, is discussed first and any necessary distinctions made later.

Evaporation occurs from both the liquid or solid phase, that is, from water or ice, provided there is an appropriate vapour pressure gradient. As pointed out earlier, there is a difference in the vapour pressures over water and ice, but if the state is known, the appropriate tables may be used.

Some of the surfaces from which water may pass are those of lakes, rivers, ponds, cloud particles, rain drops, snow, frost, moist soil, animals, and plants. Each, by its structure, may impose restrictions on the internal movement of water, but at the interface between the water and air, there is no fundamental difference and the laws of physics can be applied.

9.2 The physics of evaporation. For evaporation to occur, it is necessary to have:

(*a*) *A supply of water.*

(*b*) *A source of heat* which may be direct solar energy, R_c, sensible heat from a stream of air, H, heat from the ground, G, or stored heat from the water itself, R_s.

(*c*) *A gradient of concentration*, $e_s - e_d$. When a positive gradient exists, evaporation or moisture transfer occurs, and the necessary latent heat is withdrawn from one of the sources mentioned above. This leads to a readjustment of temperature and vapour pressure gradients with the factors reaching a balance.

Evaporation is studied from the point of view of the energy balance or through its vapour flow flux or aerodynamics. These are considered in turn and are illustrated by Figs. 9.1 and 9.2.

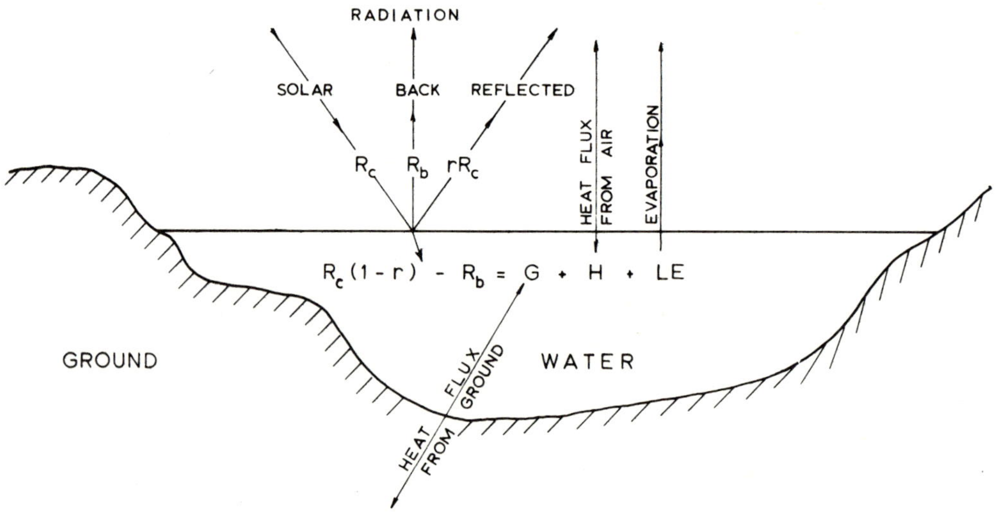

Fig. 9.1. Surface energy balance.

9.3 The energy balance approach. The energy balance approach requires knowledge, not of the process of vapour transfer, but of the factors contributing to the thermal balance at the evaporating surface. These are,

R_c the incoming radiation which may be direct, scattered or reflected short wave and long wave from sky and clouds.

$R_c\,(1-r)$ is incoming radiation into a surface of albedo 'r'.

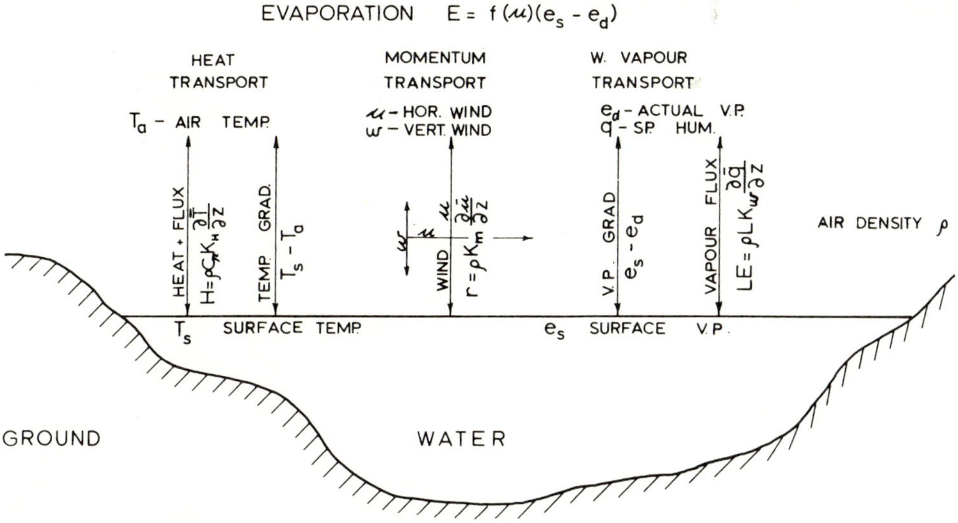

Fig. 9.2. Vapour flow.

R_b is the outgoing radiation from the surface and may be reflected short wave or long wave.

H is the sensible heat transfer from air to surface or in the opposite direction.

LE is the heat used in converting liquid to vapour, L being the latent heat and E, the evaporation.

G is the heat flux into ground or vegetation or in the opposite direction.

R_s is the heat stored in the water.

R_p is the heat converted to chemical energy in the process of photo-synthesis.

R_i is the heat moved into or out of the system as by water inflow or outflow.

R_n is the net radiation received by the surface and $R_n = R_c\,(1 - r) - R_b$.

At the evaporating surface, the principle of conservation of energy gives

$$R_n = H + LE + G + R_s + R_p + R_i \tag{9.1}$$

or

$$R_n = H + LE + G \text{ neglecting storage terms} \tag{9.1a}$$

Over short periods, R_s, R_p and R_i are not usually significant and may be disregarded. However, R_s and R_i can be allowed for from knowledge of the temperatures and volumes of the water in question, and, instead of R_n, the factor $R_n - R_s - R_i$ may be used in expressions which follow. R_p is rarely taken into account as it is usually less than 1% for most plant surfaces. The simple surface energy balance is shown by Fig. 9.1.

(*a*) *Net radiation, R_n, measurement and estimation.* Data on net radiation is best obtained by direct measurement from radiometers which have already been described. However, this type of data is not widely available so it is frequently necessary to use empirical relationships to estimate R_n.

An acceptable relationship is

$$R_n = R_a\left(a + b\,\frac{n}{N}\right)(1 - r) - \sigma\, T_a^4\,(0{\cdot}56 - 0{\cdot}09\,\sqrt{(e_d)})\left(0{\cdot}1 + 0{\cdot}9\,\frac{n}{N}\right). \tag{9.2}$$

Here a and b vary with latitude and season but can be approximated in Table 9.1.

Table 9.1. *Values of a and b in relation* $\left(a + b\frac{n}{N}\right)$.

Values	a	b
Penman at Rothamsted	0·18	0·55
Prescott at Canberra	0·25	0·54
Scholte Ubing (World data)	0·29	0·54

Also, the constants for back radiation have a range of values found by various workers but differences are small and the values given are by Penman. In this equation e_d is in millimetres.

R_a in cal/cm²/day are given in Fig. 2.3 and Table 2.4. Values of σT_a^4 in cal/cm²/day are given in Fig. 2.5 and Table 2.5. Radiation is converted to equivalent evaporation depth using the relationship, 59 cal/cm² evaporates 1 mm of water.

(*b*) *Heat flux into or out of the ground, G.* Ground heat flux, G, can be neglected if short periods of time are involved. If it is necessary to measure ground heat flux, this can be done by heat flux plates. These are similar to the elements of a radiometer and are buried flat in the ground. Temperature differences across the faces give a measure of the heat flow perpendicular to the plates.

Heat ground flux may also be deduced from changes in the temperature profile of the soil or water, together with measurements or estimates of its heat capacity.

(*c*) *Sensible heat flux and water vapour flux, H.* Sensible heat, H, or the transfer of heat between the air and water surface is not easily measured directly. It presents the same problems in measurement as the transfer of water vapour between the air and water surface or the evaporation, E.

If the transfer of heat and water vapour are considered similar processes, and transfer coefficients, K_H and K_W, appropriate to laminar flow assumed, then Bowen [1] showed that heat and water transfer are proportional and established Bowen's Ratio

$$\beta = \frac{H}{LE} = 6 \times 10^{-4}\,p\,.\,\frac{T_s - T_a}{e_s - e_d}\,. \tag{9.3}$$

Here temperature is in °C, p and e are in the same units. This means there is a constant division of the available energy between the transfer of heat and the transfer of water vapour. The value of β has been confirmed by other studies including these at Lake Hefner. [2,3]

Bowen's ratio may also be expressed in terms of the temperature and specific humidity gradients, namely

$$\frac{\partial T}{\partial z} \quad \text{and} \quad \frac{\partial q}{\partial z}$$

and using appropriate transfer coefficients K_H and K_W,

$$\beta = \frac{H}{LE} = \frac{\rho C_p K_H \dfrac{\partial T}{\partial z}}{\rho L K_W \dfrac{\partial q}{\partial z}} = \gamma \frac{\Delta T}{\Delta q} \tag{9.4}$$

where C_p = specific heat at const. pressure.

$$\gamma = \frac{C_p}{L} = 4{\cdot}2 \times 10^{-4}/^\circ\mathrm{C},$$

a psychrometric constant in terms of specific humidity (q).

Δq and ΔT are specific humidity and temperature differences measured over the same height interval. Under conditions of strong heating it was found that, about 1·5 m above the ground, due to convection, Bowen's ratio and the ratio of K_H to K_W could vary considerably. However, closer to the ground between the laminar and convective layers, equal transfer coefficients could be assumed with constant flux.

The value of Bowen's ratio changes if the surface transferring heat differs in area from that transferring water vapour. For example, studies on the U.S. class A land pan [3] showed that the top surface only evaporated, although all surfaces conducted heat to the air, the ratio of H to LE for this pan was 2½ times the accepted value. For lakes however, the usual value is appropriate.

9.4 Calculating evaporation by the energy balance approach. If G is neglected or included with heat storage, R_s, and heat inflow R_i in R so that $R_n - G - R_s - R_i$, the heat balance equation simplifies to

$$R = H + LE.$$

Using Bowen ratio

$$\beta = \frac{H}{LE} = \frac{T_s - T_a}{e_s - e_d}$$

$$E = \frac{R}{L\,(1 + \beta)} \tag{9.5}$$

$$E = \frac{R}{L}\,\frac{1}{1 + \gamma\left(\dfrac{T_s - T_a}{e_s - e_d}\right)} \tag{9.6}$$

R can be determined from radiometer and flux plate measurements or by estimation from the appropriate formulae. Bowen's ratio, β, may be determined either from equation 9.3 or 9.4 above, or from the principles of wet bulb hygrometry, measurements of temperature and humidity at two levels being required.

9.5 Vapour flow approach. It was early realised, by Dalton, and others, that the transfer of vapour must occur along a gradient of moisture concentration at a rate dependent on turbulent mixing, and a measure of the transfer coefficient might be a function of wind speed.

i.e.
$$E = f(u)\,(e_s - e_d) \tag{9.7}$$

where $f(u)$ is a function of wind speed which can be empirically derived. It is commonly of the form $f(u) = a(1 + bu)$.

The evaporation is expressed as a rate over the period in which $e_s - e_d$ and u are measured, e_s is measured at the evaporating surface, e_d in the free air, usually at 1 m, and u is an average wind speed at a suitable height. The constants a and b are fixed according to the heights selected for e_d, and u, and the nature of the evaporating surface.

Some of the more common formulae for E and e, in the same units, are

Rohwer $E = 0{\cdot}40\,(1 + 0{\cdot}27u_0)(e_s - e_d)\,u_0$ in miles/hr at surface.

Penman (1948)

$$E = 0{\cdot}40\,(1 + 0{\cdot}17u_2)(e_s - e_d),\ u_2 \text{ in miles/hr at 2 m.}$$

$$E = 0{\cdot}35\,(1 + 0{\cdot}98 \times 10^{-2}\,u_2)(e_s - e_d),\ u_2 \text{ in miles/day at 2 m.}$$

$$E = 0{\cdot}35\left(1 + \frac{u_2}{100}\right)(e_s - e_d)$$

Penman (1956)

$$E = 0{\cdot}35\left(0{\cdot}5 + \frac{u_2}{100}\right)(e_s - e_d). \tag{9.8}$$

Lake Hefner Class A pan,

$$E = (0{\cdot}42 + 0{\cdot}0040u_p)(e_s - e_d),\ u_p \text{ in miles/day at height 6 in. above}$$

rim a closer fit is

$$E = (0{\cdot}37 + 0{\cdot}0041u_p)(e_s - e_d)^{0{\cdot}88}. \tag{9.9}$$

Lake Hefner for Lake,

$$E = (0{\cdot}00304u_4)(e_s - e_2),\ e_2 \text{ at 2 m, } u_4 \text{ in miles/day at 4 m.} \tag{9.10}$$

Lake Hefner for Lake,

$$E = \frac{1{\cdot}95}{p}\,u_4\,(e_s - e_4).\ e_4 \text{ and } u_4 \text{ at 4 m.}$$

$$E = \frac{2{\cdot}3}{p}\,u_2\,(e_s - e_2),\ e_2 \text{ and } u_2 \text{ at 2 m.}$$

Here p is the barometric pressure in the same units as e. E in the last two formulae is in mm/day. In the other formulae E is in same units as e. Wind may be adjusted for height by using the logarithmic or power profile discussed in 6.4.

Empirical equations of the above type for evaporation give good results as the constants are derived from long term averages and the vapour pressure differences and wind velocities can be measured reasonably well. The main difficulty is in securing readings of surface vapour pressure or water temperatures as it is not usual practice to make these observations. However, if estimates of evaporation from reservoirs are required, it is possible to make such measurements and apply the vapour flow equations. A few years of observations enable assessments to be made of evaporation at the precise site in question and is preferred to indirect or deduced information.

Although the constants of the Dalton type equations are known for water surfaces, they have not been derived for bare soil or vegetated surfaces. This limits the application of the equations, although it is possible to use crop or soil factors.

The fact that surface properties are not known have led some workers, especially in agriculture, to use similar type formulae with the saturation vapour pressure of the air,

e_a, instead of e_s saturation vapour pressure of the surface, and to relate evaporation to saturation deficit, $e_a - e_d$. A formula of this type is

$$E = 21{\cdot}2\ (e_a - e_d). \qquad (9.11)$$

Here E and $e_a - e_d$ are in the same units and represent monthly values of evaporation and saturation vapour pressure deficit. It is obvious that this does not represent the vapour pressure difference between water and air unless the temperature of the water is the same as that of the air over the period. Although this may coincide for certain periods or occasions, that is accidental and there are as frequent divergences.

9.6 Combination of energy balance and vapour flow approaches. The deficiencies of data concerning the evaporating surfaces leads to difficulties in using both the surface energy and the vapour flow approaches. A brief experiment may give sufficient information for useful estimates at the particular site. However, it is possible, by making assumptions which do not involve much error, to eliminate surface properties by combining the formulae from the two approaches. The commonly measured meteorological elements can then be used to estimate the evaporation for past, present, or seasonal conditions. Using equation 9.7, $E = f(u)\,(e_s - e_d)$ and the saturation deficit, $e_a - e_d$, we obtain

$$E_a = f(u)\,(e_a - e_d) \qquad (9.12)$$

eliminate $f(u)$ by dividing 9.12 by 9.7 and we have

$$\frac{E_a}{E} = \frac{e_a - e_d}{e_s - e_d} = 1 - \frac{e_s - e_a}{e_s - e_d}$$

Referring to the curve of saturated vapour pressure with temperature Fig. 9.3, it can be seen that

$$\frac{e_s - e_a}{T_s - T_a} \quad \text{or} \quad \frac{e_a - e_s}{T_a - T_s}$$

approximately equals the slope of the $e:T$ curve at T_a, or,

$$\Delta = \frac{e_s - e_a}{T_s - T_a}$$

Using this in

$$\frac{E_a}{E} = 1 - \Delta\,\frac{T_s - T_a}{e_s - e_d}$$

viz.

$$E = \frac{R/L}{1 + \gamma\,\dfrac{T_s - T_a}{e_s - e_d}}$$

combine this with equation 9.6 to eliminate

$$\frac{T_s - T_a}{e_s - e_d}$$

we get

$$E = \frac{R/L + \dfrac{E_a\gamma}{\Delta}}{1 + \dfrac{\gamma}{\Delta}} = \frac{\dfrac{R}{L}\Delta + E_a\gamma}{\Delta + \gamma}. \tag{9.13}$$

Values of Δ, or the slope of the $e:T$ curve, measured in mm per degree Fahrenheit, are shown plotted against temperature in Fig. 9.4. If Δ is used in these units corresponding units for γ must also be used. This is 0·27 mm per degree Fahrenheit.

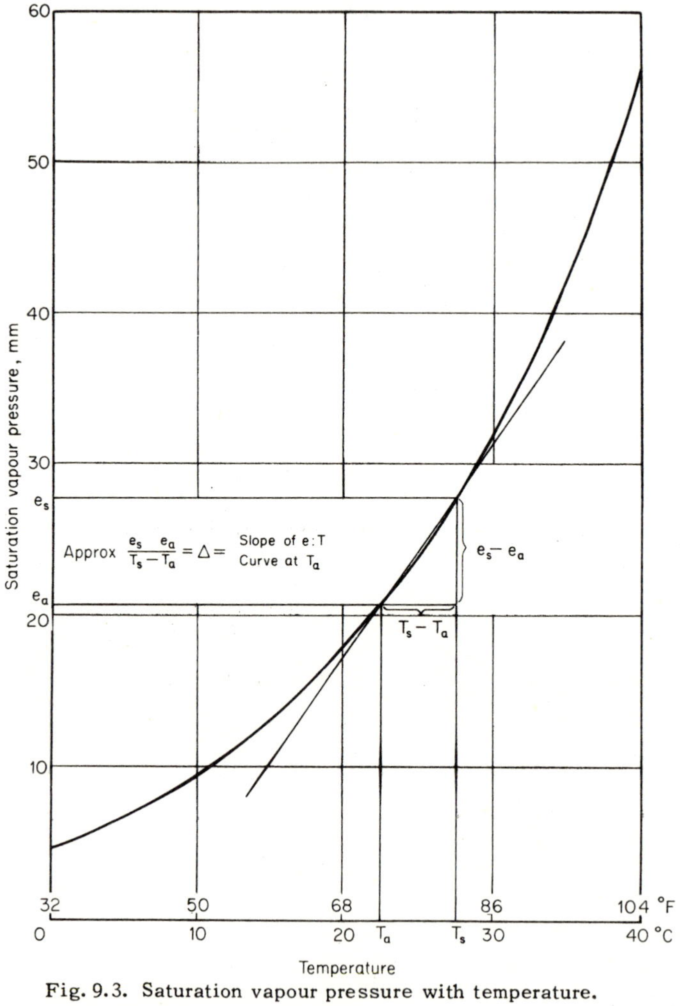

Fig. 9.3. Saturation vapour pressure with temperature.

From these and allied expressions, evaporation can be evaluated from readily available meteorological data without using the properties of the surface.

These estimates are most reliable when measured values of net radiation are used, but useful approximations are obtained when the radiation is estimated from extra-terrestrial solar radiation, sunshine, temperature and humidity data.

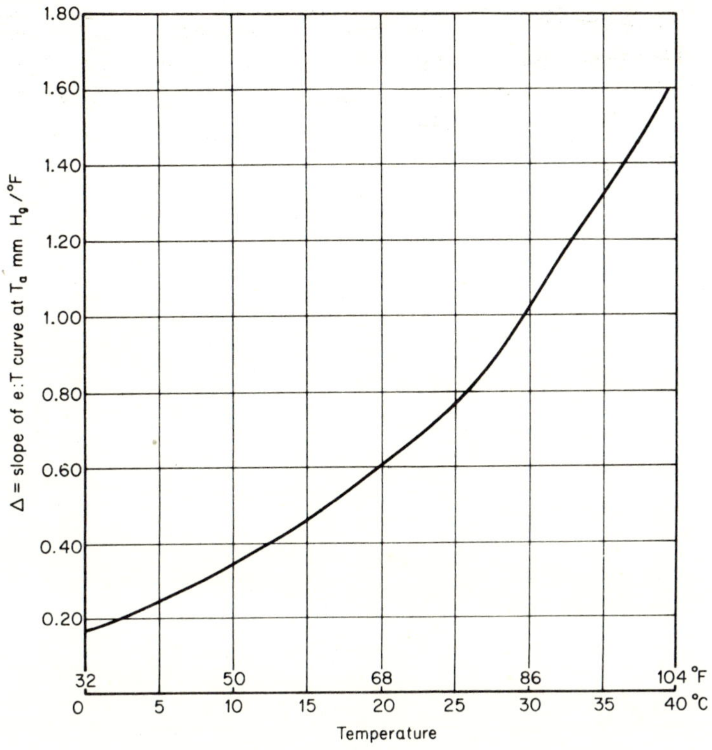

Fig. 9.4. Slope of $e{:}T$ curve:- Δ varying with T_a.

A computation of evaporation from commonly available meteorological data using the combined approach is as follows:

At latitude 45°N on July 15.

Air temperature $T_a = 68°\text{F}$ or $20°\text{C}$.

Actual vapour pressure $e_d = 13{\cdot}25$ mm.

Saturation vapour pressure $e_a = 17{\cdot}53$ mm.

(from observations $T_a = 68°\text{F}$, $T_w = 62°\text{F}$, $T_d = 60°\text{F}$ or $R.H. = 70\%$, and tables or nonograms).

Sunshine, actual hours $n = 7{\cdot}6$ hr (observed)

total possible, $N = 15{\cdot}2$ hr (from tables)

Wind, average, miles/day at 2m, $u_2 = 120$.

(observed and adjusted to 2 m if necessary)

Radiation by calculation.

solar radiation $R_a = 980$ cal/day $= 16{\cdot}6$ mm/day evap.

(from latitude, 45°N, time of year, July 15, and tables or Fig. 2.3)

at earth's surface water $R_c = R_a\,(1 - r)\left(0{\cdot}18 + 0{\cdot}55\,\dfrac{n}{N}\right)$

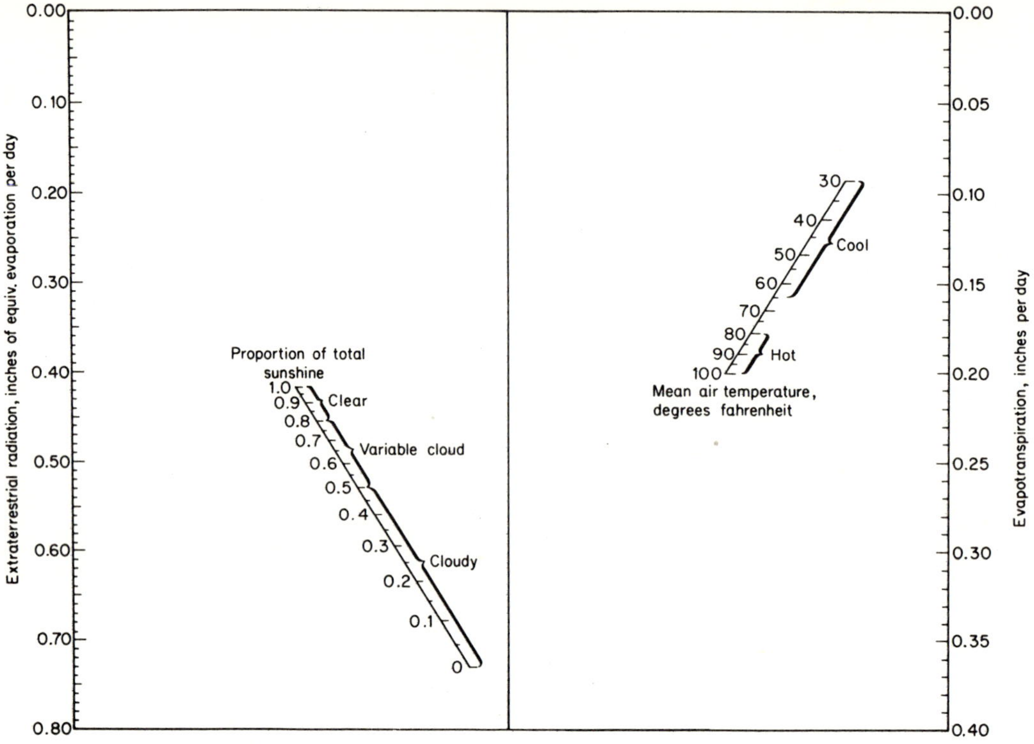

Fig. 9.5. Simple nomogram for calculating evaporation (Van Bavel).

assume albedo $r = 0{\cdot}05$ for open water

$$R_c = 16{\cdot}6,\ 0{\cdot}95\left(0{\cdot}18 + 0{\cdot}55\ \frac{7{\cdot}6}{15{\cdot}2}\right)$$

$$= 7{\cdot}17\ \text{mm/day or } 423\ \text{cal/day.}$$

Terrestrial radiation $\quad R_b = \sigma T_a^4\ (0{\cdot}56 - 0{\cdot}09\ \sqrt{(e_d)})\left(0{\cdot}1 + 0{\cdot}9\ \frac{n}{N}\right)$

(from tables or Fig. 2.5 $\sigma T_a^4 = 893$ cal/day)

$$R_b = 14{\cdot}6\ (0{\cdot}56 - 0{\cdot}09\ \sqrt{(13{\cdot}25)})\left(0{\cdot}1 + 0{\cdot}9\ \frac{7{\cdot}6}{15{\cdot}2}\right)$$

$$= 14{\cdot}6.\ 0{\cdot}23.\ 0{\cdot}55$$

$$= 1{\cdot}85\ \text{mm/day} = 109\ \text{cal/day.}$$

Net radiation $\quad R_n = R_c - R_b = 314$ cal/day

$$= 7{\cdot}17 - 1{\cdot}85$$

$$= 5{\cdot}32\ \text{mm/day.}$$

Radiation by observation (instruments).

net $R_n = 300$ cal/cm^2/day.

Equivalent evaporation $= 5{\cdot}1$ mm/day.

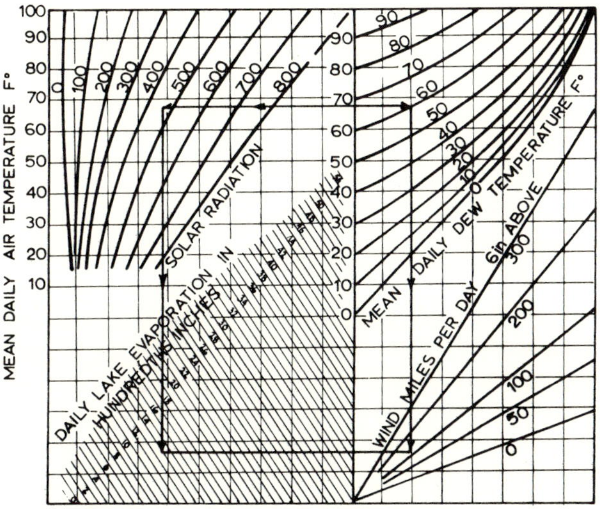

Fig. 9.6. Evaporation nomogram.

Computing E_a by vapour flow equation 9.8,

$$E_a = 0\cdot35\ (e_a - e_d)\left(0\cdot5 + \frac{u_2}{100}\right)$$

$$= 0\cdot35\ (17\cdot53 - 13\cdot25)\left(0\cdot5 + \frac{120}{100}\right)$$

$$= 0\cdot35\ \ 4\cdot28\ \ 1\cdot7$$

$$= 2\cdot55\ \text{mm/day}.$$

Combining vapour flow, E_a, and energy balance R_n,

$$E = \frac{\Delta R_n + 0\cdot27\,E_a}{\Delta + 0\cdot27}$$

$$= \frac{0\cdot61.\ 5\cdot32 + 0\cdot27.\ 2\cdot55}{0\cdot61 + 0\cdot27} \quad \text{where } \Delta \text{ at } 68^\circ\text{F} = 0\cdot61.$$

$$= 4\cdot47\ \text{mm/day}.$$

Computer programmes have been prepared for the rapid evaluations of E. [4] In addition nomograms give a quick method of estimation. A simple one adapted from Van Bavel [5] is given in Fig. 9.5. A more elaborate one following the work at Lake Hefner is shown in Fig. 9.6. [6] From this diagram, if air temperature $T_a = 68^\circ$F, dew point $T_d = 60^\circ$F, solar radiation = 460 cal/cm^2/day and wind at 6 in. above pan or 22 in. above ground = 120 miles/day, then open water evaporation is 0·14 in./day = approx 3·6 mm/day.

9.7 Vapour flow (aerodynamic) approach. If evaporation is considered as the turbulent transfer of water vapour at a small distance above the ground, the empirical Dalton coefficient, $f(u)$, can be replaced by the product of air density and the eddy transfer coefficient, ρK_w. The surface to air vapour pressure difference, $e_s - e_d$, can also be replaced by the average humidity gradient, $\partial\bar{q}/\partial z$.

That is,
$$E = -\rho K_w \frac{\partial \bar{q}}{\partial z}, \tag{9.14}$$

similarly the transport of heat,
$$H = -\rho C_p K_H \frac{\partial T}{\partial z}, \tag{9.15}$$

and the transport of momentum,
$$\tau = \rho K_m \frac{\partial \bar{u}}{\partial z} \tag{9.16}$$

where K_H and K_m are eddy transfer coefficients for heat and momentum.

With careful instrumentation $\partial\bar{q}/\partial z$ can be measured but K_w is not easily expressed in terms of other measureable quantities. It is reasonable and convenient to assume K_w equal to K_m, which is more easily determined from the changes of wind speed with height. A commonly accepted variation is the logarithmic profile which gives for measurement of q and u at the two levels, z_1 and z_2.

$$E = \frac{k^2 (q_1 - q_2)(u_2 - u_1)}{\log_n \left(\frac{z_2}{z_1}\right)^2} \tag{9.17}$$

where k is von Karman's const. $= 0{\cdot}4$.

This was tested, mostly over open grassland, and can be re-expressed in a form similar to Dalton's equation.

$$E = B\, u_2\, (q_1 - q_2) \tag{9.18}$$

where

$$B = \frac{k^2 m}{R\, T_a} \; \frac{1 - \frac{u_1}{u_2}}{\log_n \left(\frac{z_2}{z_1}\right)^2} \tag{9.19}$$

m being molecular weight of water.

Under near neutral conditions u_1/u_2 depends on the roughness of the surface and B is therefore nearly constant for surfaces where roughness remains the same. Therefore if B is first determined from a set of careful difference measurements, then subsequent calculations of E require wind speeds at only one level. Atmospheric stability affects the relationship between K_w and K_m unless observations are made close to the evaporating surface or another wind profile is used to determine K_m, for example, the one seventh power wind profile.

9.8 Direct eddy flux approach. A fundamental approach is to measure the actual turbulent air movements, ρw, which are part of the eddy diffusion process, and the associated humidities, q. This permits direct evaluation of the eddy flux of water vapour, that is, ρwq, and is independent of the nature of the surface and of the atmospheric conditions.

Sensitive anenometers together with wet and dry bulb thermometers have been used to measure the required elements and the computation is achieved by an analyser. [7] This experimental approach promises a method of determining actual evaporation or water loss from a surface irrespective of its nature.

9.9 The hydrological balance. The various parts of the hydrological cycle shown in Fig. 1.1, can be expressed in the form of an equation, bearing in mind the principle of conservation of mass or the principle of continuity. Thus the inflow or precipitation, P, equals the outflow plus the change in the system. Outflow factors are evaporation, E plus surface runoff, Q, plus subsurface drainage, D, while change factors are those in storage, ΔS.

That is $$P = E + Q + D + \Delta S, \tag{9.20}$$

both Q and D are net quantities being the difference between outflow and inflow into the part of the cycle considered. ΔS is the change in storage. This may be surface or channel storage, soil moisture storage, or ground water storage. The small percentage of water held in plant and animal life may be considered part of the surface storage and does not vary greatly in the period concerned.

By measuring or eliminating all but one of the factors in this water balance equation, the remaining one is estimated.

This direct and simple method is applied in the evaporimeter or lysimeter where precipitation, overflow and the change in the enclosed volume of water or soil are known and evaporation or transpiration therefore deduced.

The water balance equation is often applied to lakes where careful measurements of the various factors enable accurate estimates of actual evaporation to be made. [3]

The same equation is also applied to suitable catchment areas with success. By selecting times at which the catchment is at a similar state, it is possible to neglect the storage factor, or $\Delta S = 0$. Precipitation over the catchment and net runoff are easily computed and sub-surface drainage is often small enough to be neglected, or a rough estimate is not largely in error. Hence catchment evaporation can be surmised. This gives a useful check on other methods of measuring evaporation or on the overall standards of measurement in the catchment.

On a catchment, the water balance equation is most useful for long term estimates as all changes in the period are integrated. Notwithstanding this, lysimeters and tanks, considered as particular applications of the water balance equation, are used over short periods which may be a day or even hours when evaporation rates are high.

Although the data of particular catchments or tanks apply to these specific sites, they may be transferred and applied to other localities and surfaces under certain circumstances. This is further discussed later under pan coefficients and transpiration.

9.10 The control of evaporation. The prevention or restriction of evaporation is an attractive method of conservation and considerable attention has been devoted to it. Methods of control stem from the physics of evaporation, and seek to reduce the radiation or heat energy received by the water (energy balance) or to materially reduce the vapour pressure gradients and wind velocities (Dalton's equation). These methods and principles may, with advantage, be applied in both design and operation of a project.

9.11 Evaporation and the design of a project. In the design of a water project, its probable evaporation losses should be considered. For example, if there are a number of satisfactory locations available for reservoirs, probable evaporative losses for each site might be computed by the methods mentioned here. The site with the minimum losses may be more suitable on that score.

By experience or intuition, many of the principles of reducing evaporation are already being considered or used in design. For example, deep reservoirs with minimum surface areas receive less radiation per unit of water and so have lower evaporative losses. A reservoir within a steep gorge may be so orientated and sheltered that it receives less radiation and wind advection with corresponding lower evaporation than other locations. Reservoirs sited within climatological zones of low evaporation, for example, in humid or semi-humid regions where there are small vapour pressure differences, or at high altitudes with large nocturnal radiation losses, lower temperatures and small vapour pressure differences, may supply water to a nearby zone more efficiently than one located in the arid zone itself where there are high evaporative losses.

9.12 Evaporation and the operation of a project. The careful operation of a water project may reduce evaporative losses considerably. For example, the drawing off of the warmest water in the reservoir extracts heat which might otherwise be spent in evaporation. If the warmest water is at the top of a reservoir, this is frequently also the purest water.

In a complex of reservoirs, the first use of the warmest water or that from areas prone to high radiation such as shallow stretches in arid zones, may reduce the annual evaporative loss. It can be an economical method of reducing evaporation to improve the shape of a reservoir by building dykes to cut off large shallow areas which absorb and pass into storage large amounts of radiation and present broad surfaces for evaporation.

Shelter belts can be constructed to reduce the wind flow and heat advection over the water surfaces, and to modify the local vapour pressure gradients. If these belts are trees, their consumption of water is balanced against the possible reduction of evaporation. Shelter belts are of value to small reservoirs and ponds rather than larger storages.

9.13 Methods of controlling evaporation. One of the simplest methods of reducing evaporation imposes a physical barrier between the water surface and the air. Underground storages are of this type and are effective in conserving water with little evaporation. Man-made underground tanks or roofed reservoirs store small quantities of water without evaporative losses, but are expensive.

Groundwater reservoirs, at sufficient depth, have low evaporative losses, but data on their formation, structure, and operation are limited. Seepage, or underground flow, may be a greater loss than evaporation. Where a reliable ground water basin is in use or available, it can be recharged by allowing surplus water to flow into the intake beds or wells. The water may then be held in storage with small evaporative losses and withdrawn when required.

The use of a mulch over soil is an effective method of conserving water. Radiation is reflected and temperature, vapour pressure, and wind profiles are modified. The efficiency depends upon the albedo, pore space, and nature of the material and the underlying surface. Mulches of a few inches depth have reduced soil evaporation by one half with greater reductions in specific circumstances.

It has been suggested that water surfaces might be protected from evaporation by floating covers or layers of small spheres such as the froth of a polystyrol, but no economic method on these lines has yet evolved.

Another form of barrier is a thin chemical film spread over the water surface. This would have to be non volatile, and float on the surface. Many hydrocarbons come within this category, and thin oil has been commonly used in storage precipitation gauges. Such a film, however, is injurious to biological life and prevents the exchange of gases between air and water. Experiments using certain alcohols, e.g. hexadecanol, $CH_3(CH_2)_{15}OH$, or octodecanol, showed it was possible for a monomolecular layer of such material to spread over a water surface without harmful effects. [8] These particular materials have one end of their chain structure hydrophilic and the other hydrophobic and consequently arrange themselves, one molecule thick, on end, over a water surface.

The monomolecular films have reduced evaporation by various amounts dependent on the weather, with the highest reduction in mild calm conditions. The reduction has been estimated as 50% in mild weather, 40% in hot weather with low wind, and 20% or lower in hot windy weather. [9] Tests on actual reservoirs have proved a reduction of 14% at Lake Hefner and this can be accepted as a conservative figure for most conditions. It has been found impractical to maintain a coverage with wind above 15–20 miles/hr as the film is relatively inelastic and does not recover well. The material may also be attacked by bacteria.

Hexadecanol or octodecanol, may be dispensed by using pellets in wire screen floats, as a mechanical suspension, as a solution, as an emulsion, as a powder, in flakes, or sprayed on water in a molten condition. The method used depends upon the area to be serviced and the conditions at the site.

The cost of applying anti-evaporants is comparable with costs of reclaiming waste water but depends very much on local conditions, the value of water, and labour costs.

9.14 Measurement of evaporation by pans. Theoretical and empirical methods of estimating evaporation have been discussed together with some experimental methods. There are also many comparatively simple devices such as pans which purport to measure this phenomenon and, without doubt, they give the water loss from the containers used. In most cases however, it is evaporation from a larger water surface, from a catchment, from a crop, or from soil, which is required and there are difficulties in establishing the relationships between the water loss from the pans and these other quantities.
are first described and information on their relationships given later.

There are four main types of evaporimeters; floating pans, pans above ground, pans sunk in soil, and devices with special evaporating surfaces. The principal types only in each category are described briefly. If required, full details may be obtained from the relevant publications of the meteorological services concerned or from texts on instruments.

(*a*) *Floating pans.* A pan, usually of the dimensions of the A class land pan, can be made to float within a reservoir with suitable rafts. Provided the air flow is not unduly disturbed by the installation, water losses are similar to those of the lake with an allowance for the rim effect. There are considerable difficulties in operation, however, as wind and wave action and splashing make the results unreliable. Also, good installations are expensive to set up and maintain, so few results are available, except in experimental sites.

(*b*) *Above ground pans.* The class A land pan is now widely used, as much data on its relationships with other surfaces are available and it is not subject to the variable effects of heat flow from the ground. It consists of a galvanised iron pan, 4 ft in dia., 10 in. deep, and supported 6 in. above ground on a wooden frame which allows air to circulate under it. The water level is maintained at a standard 7½ in.

In the A class pan, heat flux from the ground (G) is negligible but considerably more sensible heat (H) is received through the sides and bottom of the pan than by a sunken pan or reservoir, resulting in higher evaporation. For the A class pan the Dalton relationship is

$$E_p = (e_s - e_d)^{0\cdot88}\ (0\cdot37 + 0\cdot0041u_p) \tag{9.21}$$

and Bowen's ratio for the pan is 0·025 compared with 0·0105 for a lake.

An interesting variation to avoid excess heat advection is an insulated A class pan with the rim shaded to avoid radiation other than on the water surface. This pan becomes virtually an integrator of radiation. [3]

In an attempt to maintain constant evaporation conditions, the black pan evaporimeter was suggested. It consists of a shallow black pan approximately 64 cm in dia., 2½ cm deep, supported at a height of 1 m on a wooden platform, with water maintained to a depth of about 1 cm from a suitably insulated tank. The drawdown of this tank gives the black pan evaporation which is about 25% more than for the A class land pan. Rain days introduce inconsistencies and with high wind, water is blown out of the pan with anomalous results.

(c) *Sunken pans.* Most countries have used sunken pans for many years. The United States Bureau of Plant Industry pan is 6 ft in dia., 2 ft deep and buried with the water surface at approximately ground level. A similar pan is the Colorado which is 3 ft square and 1 ft 6 in. deep. In an attempt to obtain a pan factor of unity the Young pan is 2 ft in dia., 3 ft deep and covered with ¼ in. mesh.

The British Institute of Water Engineers' standard pan is of wrought iron and is 6 ft square and 2 ft deep. It is sunk to within 3 in. of ground level at which the water surface is maintained.

An Australian pan is 3 ft in dia. and 3 ft deep, but contained in a second pan 4 ft in dia. and 2 ft 10 in. deep. It is buried with the outer rim level with the ground surface at which level water is maintained in both tanks. Measurements are taken in the inner pan.

The sunken tanks suffer from operational difficulties in that they must be kept clean of the rubbish which collects in them. Being in the microclimatological zone, the air flow, temperatures, and humidities can vary with season and surrounding vegetative cover. In arid climates in particular, considerable heat leakage occurs between the pan and the surrounding soil. All of these factors give variable pan coefficients and correlations.

Lysimeters are special forms of sunken pans where the water can be replaced with blocks of soils in which moisture conditions are varied or crops grown. They range from expensive recording balance instruments which show dewfalls and hourly transpiration rates to simple draining potential transpirimeters. In the latter, an oil drum is filled with soil in which may be grown a crop and, it is supplied daily with a measured amount of water. The drainage is measured and the water loss computed from the water balance of the soil block. [11,12]

Lysimeters, as distinct from the potential evapotranspirimeters, are capable of determining the true evaporation from most types of surfaces. They should be suitably designed and installed with appropriate guard rings of similar surfaces around them.

(*d*) *Special devices for measuring water losses.* A number of instruments, apart from pans, have been evolved in attempts to measure evaporation or transpiration. The commonest is probably the Pische evaporimeter which comprises an inverted graduated tube 22 cm long, filled with water, with a filter paper of 3·2 cm dia. clamped over the end. The whole is kept in a Stevenson screen. Evaporation occurs from the 13 cm^2 of the filter paper and is measured in mm of water lost from the tube.

Readings can be converted into depths per unit area. This instrument tends to overestimate the effect of wind and under estimate that of radiation. Consequently little consistency has been found in correlating Pische readings with surfaces differently exposed.

Other instruments in use comprise ceramic plates supplied with water and, although these may more nearly approximate conditions of leaf surfaces, water or soil, energy balance and wind conditions for the instrument and the other surfaces, differ and, there is no biological control. However, under experimental conditions useful correlations have been obtained, but it is doubted whether these can be applied in other areas.

9.15 Factors affecting pan evaporation. Experiments show the effect of diameter, rim height, colour and depth on pan evaporation. The results can be explained in terms of the physics of evaporation and are given below.

The rate of evaporation from a pan decreases as its diameter increases, but beyond a diameter of about 3½ m or 12 ft, there is little significant change in the evaporation rate. This can be explained by turbulence theory and a general formula has been derived for evaporation rate equal to

$$\frac{\partial E}{\partial x} = \text{const. } u^{0\cdot78} \, x_0^{-0\cdot11} \tag{9.22}$$

where x_0 is a strip length down wind, u.

The rim height affects wind velocities, turbulence, and the radiation which is reflected or absorbed by the portion of rim exposed and cut out by that portion shading the pan. These effects are of more significance in the smaller pans and shields have been devised to reduce their influence. If the water surface is not maintained at a standard level, rim height has an influence. For example, experiments on sunken pans showed that a drop of level of 5 cm gave 15% less evaporation.

The material and colour of the pan affects the radiation absorbed and reflected by the sides and bottom of the pan, as well as the amount radiated at night. A white pan has been observed to evaporate more in winter, but less in summer than a similar coloured pan. On an annual basis, black, grey or galvanised and white showed decreasing evaporation in that order.

The nature of the ground surrounding a sunken pan has a significant effect upon its evaporation in that heat may be transferred from soil to pan and vice versa. Sunken pans in arid areas frequently show greater evaporation (lower pan coefficients), than might be expected from computations of evaporation, because of the transfer of heat from bare ground to the tanks.

Increased depth has a lagging effect on evaporation in that more heat can be stored, with a slower seasonal rise of temperature and evaporation. The heat stored is utilised at a later stage in evaporation. The peak evaporation from a large lake will thus occur when $e_s - e_d$ is greatest and this may be as much as 6 months later than for a pan or small lake with less heat storage. These phase and amplitude differences show the difficulty of using pan coefficients on a monthly basis.

9.16 Pan coefficients. It is obvious that pans measure evaporation from their containers which are located in particular sites for specific periods. The extension of this information to other surfaces is not so obvious and is fraught with difficulties, problems, and some inconsistencies. Comparatively simple experiments however, give pan coefficients or factors which, when multiplied by pan evaporation, purport to give estimates of open water or lake evaporation. It is also possible to relate open water evaporation to water losses from vegetation, or transpiration, and to water losses from soil.

In view of the many factors which affect evaporation and the fact that these have no systematic variation between a pan and a lake surface, it is not surprising that a pan coefficient fluctuates, especially when monthly or weekly figures are considered. Some factors such as soil temperatures, have an annual cycle and an annual pan coefficient is then consistent and, if employed on this basis, gives reasonable results.

If surfaces other than water are under consideration, the actual evaporation, or water loss, depends primarily upon the moisture available as well as the other factors. Pan coefficients convert evaporation from the pan to lake evaporation. If actual water loss is required, it follows that direct measurement is necessary or another relationship taking account of the surface in question and the moisture supplied to it. From a variety of sources pan coefficients are quoted in Table 9.1. They should be used with caution and due reservations.

Table 9.1. *Coefficients giving relationship between pan, soil, or plant evaporation and open water evaporation* (E_w)

Type of pan or surface	Coefficient (k) $E_{water} = kE_{pan}$	Source	Remarks	Coefficient $\frac{1}{K}$ $E_{pan} = \frac{1}{K}E_{water}$
Class A land pan	0·7	A.S.C.E.	Range 0·60–0·82 (Annual)	1·4
Colorado buried pan	0·78	A.S.C.E.	0·75–0·86	1·3
U.S.G.S. floating	0·80	A.S.C.E.	0·70–0·82	1·2
Australian buried	0·9	C.B. of M.	Inland 0·7 Coast 0·9	1·1
half 44 gall drum on ground [13]	0·6	W.R.F.A.	Summer 0·65 Winter 0·54	1·7
above with ½ in. wire mesh	0·63	W.R.F.A.	Summer 0·70 Winter 0·59	1·59
Black pan [14]	0·56	C.S.I.R.O	variable	1·8
Pische [14]	0·8	C.S.I.R.O	variable	1·2
Pische	0·8	India	1·0 wet season 0·6 dry season	1·2
Young pan [15]	1·0	Ref. 15	variable	1·0
British Inst. of Water Engineering	0·9	B.I.W.E.	variable	1·1
U.S. Bureau of Plant Industry	0·9	B.P.I.	variable	1·1
Short grass [16]	1·2–1·6	Penman	1·2 summer 1·6 winter	0·8–0·6
Wet soil [12]	1·08	U.N.E.S.C.O.	short record at Davis, Aspendale	0·92
Grass and pasture [12]	0·82–0·96	U.N.E.S.C.O.	short record at Davis, Aspendale	1·22–1·04

REFERENCES

[1] BOWEN I.S., (1926). The ratio of heat losses by conduction and by evaporation from any water surface. *Phys. Rev.* **27**, 779–89.

[2] CUMMINGS N.W. and RICHARDSON B., (1927). Evaporation from lakes. *Phys. Rev.* **30**, 527–34.

[3] U.S. Geol. Survey (1954). *Water Loss Investigations – Lake Hefner Studies.* Paper 269.

[4] LAMOREUX W.W., (1962). Modern evaporation formulae adapted to computer use. *Monthly Weather Rev.* **90**, 26–8.

[5] VAN BAVEL C.H.M., (1957). Estimating soil moisture conditions and time for irrigation with evapotranspiration method. *U.S. Dept. of Ag. A.R.S. 41–11.*

[6] KOHLER M.A. and RICHARDS M.M., (1962). Multicapacity basin accounting for predicting runoff from storm precipitation. *J. Geophys. Res.* **67**, 5189–97.

[7] TAYLOR R.J. and DYER A.J., (1958). An instrument for measuring evaporation from natural surfaces. *Nature, Lond.* **181**, 408–9.

[8] MANSFIELD W.W., (1955). Influence of monolayers on the evaporation of water. *Nature, Lond.* **175**, 247.

[9] VINES R.G., (1962). Evaporation control. A method of treating large areas in *Retardation of Evaporation by Monolayers.* Ed. LA MER V.K., Academic Press N.Y.

[10] FRANZINI J.B., (1961). Evaporation suppression research *Water & Sewage Works.* **108**, 167–172, 221–225.

[11] MATHER J.R. and GARNIER B.T., (1952). A simple apparatus for measuring potential evapotranspiration. *Nature, Lond.* **170**, No. 4320.

[12] SLATYER R.O. and McILROY I.C., (1961). *Practical Microclimatology* U.N.E.S.C.O.

[13] WIESNER C.J., (1964). Water Requirements of Crops. *Report No. 9. Water Research Foundation of Australia,* Sydney.

[14] FLEMING P.M., (1962). Evaporimeter relationship at Griffith. N.S.W. *C.S.I.R.O., Griffith, N.S.W.*

[15] YOUNG A.A., (1947). Some recent evaporation investigations. *Trans. Am. Geophys. Union* **28**, 279–284.

[16] PENMAN H.L., (1948). Natural evaporation from open water bare soil and grass. *Proc. R. Soc. Lond., A* **193**, 120–45.

OTHER REFERENCES

DEACON E.L., PRIESTLEY C.H.B. and SWINBANK W.C., (1958). Evaporation and the water balance. *U.N.E.S.C.O. Review of Research* **10**, 9–34.

FERGUSON J., (1952). The rate of natural evaporation from shallow ponds. *Austn. J. Soc. Res., A.* **5**, No. 2, 315–30.

PENMAN H.L., (1950). The water balance of the Stour catchment area. *L. Inst. Water Engrs* **4**, 457–469.

PRIESTLEY C.H.B., (1959). *Turbulent Transfer in the Lower Atmosphere.* The University of Chicago Press, Chicago.

KOHLER M.A., NORDENSON T.J. and FOX W.E., (1955). Evaporation from pans and lakes. *Res. Pap. 38., U.S. Weather Bureau.*

SUTTON O.G., (1953). *Micrometeorology,* McGraw Hill, N.Y.

CHAPTER TEN

Precipitation and its relationship with time

10.1 Definition and forms of precipitation. Precipitation is the depositing of water from the atmosphere on to a surface. This deposit may be either liquid or solid to give the various forms of precipitation. Dew, fog, cloud, and frost contribute to precipitation when saturated air comes into contact with cooler surfaces, or is intercepted by vegetation or other objects. The relatively small amounts received are best measured by dew gauges which depend on the appearance of drops on specially prepared surfaces, by sensitive weighing balances, and special interception gauges. The standard rain gauges collect only small amounts of these forms of precipitation. Although not significant elements in runoff, the amounts of dew etc. reaching the ground or vegetated surfaces are sometimes appreciable and important in plant growth and development. In certain areas, shallow insulated ponds on the top of hills, known as dew ponds, cool rapidly at night and are replenished by these condensation products.

The precipitation forms of significance in surface water hydrology are drizzle, rain, snow, sleet, and hail. They result from precipitation mechanisms, described in later chapters, acting upon streams of moist air.

Liquid precipitation is classified according to the size of the droplets with drizzle generally being below 0·02 in. (0·5 mm) dia. The largest drops which can exist without breaking up are about 0·2 in. (5 mm) dia. The classification of rain as light, moderate and heavy, depends upon drop sizes, although the varied vertical velocities above a given area give a range of drop sizes for each of these forms of rain with a predominance of large drops in the heavier rainfalls. Figure 10.1 shows the distribution of drop sizes for three types of rainfall.

Snow crystals form when vapour condenses, by sublimation, to ice, below 32° F. They collect in groups as snow flakes and, being of low density, are carried freely, fall slowly, and are affected by wind and turbulence. Their measurement by gauge is thus difficult and their accumulation on the ground varied. When raindrops freeze, sleet is formed. Hail results from the freezing and melting of water droplets and ice as they are carried up and down in the turbulent currents of large clouds. Hailstones may exceed 2 in. (5 cm) dia. and may weigh up to 7 lb (3 kg). They display a banded structure consequent on the alternative melting and freezing of the layers during their descent and ascent.

10.2 Classification of precipitation. Besides classifying precipitation with respect to form and intensity, it is also classified according to the type of lifting mechanism which induces condensation and aggregation. Orographic, frontal, convective, and convergent precipitation are four important types which are described in chapter 14. The properties of the air masses associated with the lifting mechanisms, particularly their stability and moisture contents, are also significant in the precipitation rates received and qualify the above classification. For example, stable orographical rainfall implies that the air mass is stable and lifted only to the height of the mountain range. Completely unstable orographical rainfall implies that the air mass becomes unstable when lifted and may ascend to the limit of convection. The range acts as a trigger to the latent instability of the air mass.

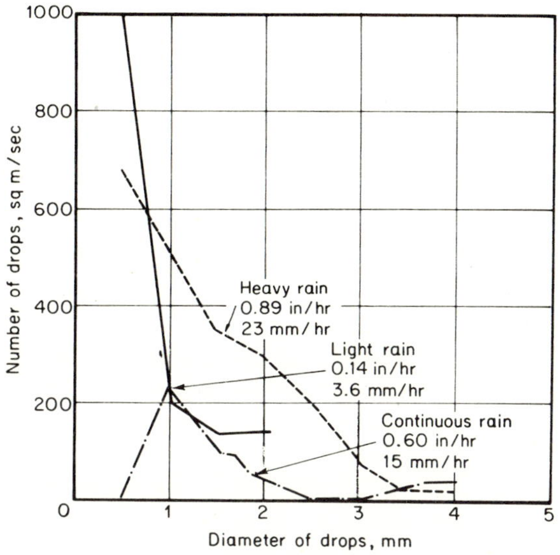

Fig. 10.1. Distribution of drop sizes in rainfall.

The precipitation intensities which result from frontal, convective or convergent action are also affected similarly by the stability of the air mass involved. It is sometimes difficult to use this classification, due to the presence of more than one air mass, or lifting mechanism, and interaction between the processes at the one location. However, components may be separated out when there are stations not influenced by one process. For example, in some situations the comparison of rainfalls on a plain with those at nearby stations on the slopes permits the determination of the orographical component.

10.3 The measurement of precipitation. The amount of precipitation received by a given area depends upon the nature of the surface and the associated wind field as well as the meteorological factors discussed in chapter 14. A water surface, being flat except for waves and swell, does not disturb the wind flow above it and receives most of the precipitation falling on it, allowing for the inclination of the raindrops' trajectory to the surface. A ground surface, on the other hand, displays numerous irregularities and slopes, both natural and artificial. These and the mantle of vegetation affect the wind flow to give varying velocities and increased turbulence. These effects, added to interception and splash, cause the ground surface to receive amounts different from the rain gauge placed according to the standard specifications. The significant factors are the inclination of the receiving surface to the falling condensation products and the wind field. The wind changes the trajectory of the falling precipitation from the vertical and more will fall on a given area at right angles to the trajectory than on the same area in a horizontal plane. The factor of proportionality between rainfall on the horizontal plane and on a given ground slope depends upon trigonometrical functions of the angle of ground slope and the angle of inclination of the raindrops' trajectories to the vertical. A network seeking to measure rainfall on the ground surface faces these difficulties of allowing for land slope and the wind field. However, if these limitations are realised, the measurement of precipitation by a depth on a horizontal surface which is suitably located, as in the standard raingauge, is a reasonable estimate of that on the nearby ground surface. Because a catchment area is computed as the projection upon a horizontal plane, the horizontal rain gauge gives a representative measurement.

In using gauges to evaluate the precipitation over an area, the following sources of error occur:

(*a*) Deficiencies in the collecting instrument.

(*b*) Unsatisfactory siting, or difficulties in locating gauges so that they adequately represent an area.

(*c*) Insufficient or poorly planned networks.

Instrument errors include the size and mode of construction of the gauge. Gauges of diameter below five in. are less accurate than the larger ones, although accuracy of construction and provision of a rim and funnel to minimise splashout and evaporation are of greater importance than actual diameter. The disturbance of the local wind flow with increased turbulence and updraft over the opening of the gauge affects the catch. Errors from this source are greatest with strong winds in exposed sites, and may become very large when gauges are used for snowfall measurements. These may be remedied to some extent by the use of shields which reduce turbulence over the gauge. [1, 1a] Nipher, Alter, and Shasta shields are examples of the devices necessary if gauges are to record snowfalls in areas of strong winds. Without shields, errors of 20% have occurred at wind speeds of 10 knots and 50% at wind speeds of 40 knots for rain, while for snow, errors of 40% and 70% have been known at these velocities. However, these are adverse conditions and exposures.

The careful siting of a gauge reduces turbulence above it, and aims at giving an air flow above the gauge similar to the surrounding area which it seeks to represent by its readings. The raising of the gauge above ground level, necessary to avoid splash from the surrounds, flooding, and accidents, introduces a disturbance of the air flow which can be remedied by shields, or by locating the gauge in a slight hollow or within a circular low wall of turf, 5 ft from the gauge. One criterion of exposure specifies that the gauge be placed one foot above the ground surface at a distance of twice the height of an obstruction from it. This gives a free line of sight above an angle of 26° from the gauge. Some meteorological services specify an angle of 45°. An excessively exposed site results in turbulence and a catch not necessarily representative of the surrounding area. Care in siting precipitation gauges and assuring that the standard is maintained during the length of record is of prime importance in obtaining reliable precipitation data. Investigations on different types of gauges show that although higher amounts of rainfall, up to 10% more, may be recorded in stereo-gauges which are cut at an angle to the horizontal, or in gauges at ground level, the observations from gauges placed in standard conditions give consistent readings which are immediately comparable and less subject to error or accidents. [2] On a larger scale, topography, vegetation and artificial obstructions disturb the wind flow, cause many variations in the precipitation pattern over an area, and introduce problems of locating gauges to adequately describe the catchment. Local knowledge of the rainfall distribution and regime from isohyetal maps and existing records, combined with personal judgement, is used in locating gauges. There is merit in using a dense network of gauges for a short period or in using refined methods such as the isohyetal technique and in comparing these results with the arithmetic average of a smaller and more convenient network. This network may be selected to give a result similar to the dense network evaluated by isohyets. For example, the arithmetic mean of eight rain gauge stations in the Macleay River catchment, N.S.W., of area 4000 square miles was found to be an acceptable approximation to the mean catchment rainfall determined by subjective isohyets from a network of all available stations, which exceeded twenty for eighteen storms. [3]

Errors in estimating areal rainfall from a given network also occur because of the random nature of storms and their passage between the gauges. A statistical analysis of the rain gauge data in a region of relatively flat topography established the relationship between the raingauge network density, the area considered, and the percentage standard error of estimation and is shown in Fig. 10.2. [4] This shows that, for the same level of accuracy, a denser network is necessary for small areas than for larger areas. For example, over 500 square miles to estimate rainfall to 15%, 1 gauge in 100 square miles is needed whereas over 8000 square miles, one gauge in 600 square miles would suffice. Figure 10.2 emphasises the need for denser networks if higher accuracy of estimation of precipitation is to be achieved.

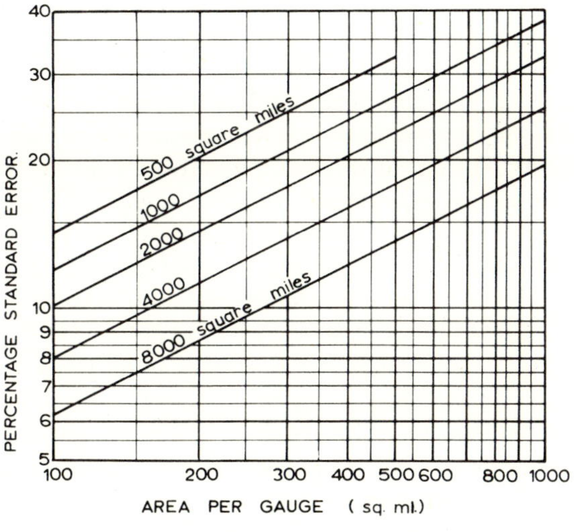

Fig. 10.2. Rain-gauge network and percentage standard error.

To determine a suitable network density many other factors have to be taken into account. The topography, climate, and state of development of a region are relevant as well as finance and the availability of suitable observers. The World Meteorological Organisation and others have set standards for optimum and minimum networks which serve as a guide.[5] The minimum network considers some of the above factors, and is given in Table 10.1.

Table 10.1. *Minimum density of precipitation network.*

Type of region	*Normal tolerance (area for 1 station)*	*Tolerance under difficulties (area for 1 station)*
Flat areas of temperate, mediterranean and, tropical zones.	600–900 km^2 230–350 square miles	900–3000 km^2 350–1160 square miles
Mountainous regions as above	100–250 km^2 39–100 square miles	250–2000 km^2 100– 770 square miles
Small mountainous islands with irregular precipitation	25 km^2 10 square miles	
Arid and polar zones.	1500–10 000 km^2 580– 3 860 square miles	

Optimum networks are more extensive and depend principally on the type of region. They should permit the determination of the average and extreme precipitation at all points by interpolation from the network. All suitable networks contain permanent stations which give the bases for statistical studies and which are observed continuously and indefinitely. Any minimum network set up should have regard to the permanent stations and their importance. Secondary stations perform observations for a limited period of time in order to establish a good correlation between themselves, the base stations, and the terrain. Special stations are those set up for specific purposes such as research projects, the control of irrigation, or intensive regional development.

To each precipitation depth corresponds a time period which may be a year, month, day, hour, or minute. Thus the element of rainfall intensity is implicit in all readings and the period is made apparent by the term annual, monthly, daily, or hourly precipitation. However, it is convenient to express intensity in terms of the rate per hour and, unless otherwise specified, the term rainfall intensity implies the rate in in./hr or mm/hr. It follows that different rates occur over different periods of time and in different storms. The more intense rates are over short periods and, as the period extends, the rate over that period declines. The measurement of these rates is an important and necessary task.

Precipitation rates over a period of a day or multiples of days are derived from daily readings of the standard gauge. Over longer periods storage gauges are used. For intensities over shorter periods and for the detailed storm pattern, readings must be taken at smaller intervals. This is achieved by means of pluviographs. Details of the many types of recording precipitation gauges are obtainable from books on instruments or from the national meteorological services. Their location and siting follow similar principles to those of the rain gauge and demand the greatest of care. From the pluviograph's record of rainfall depth against time, intensities can be computed over the desired periods and hyetographs and mass curves of rainfall constructed. The degree of precision achieved depends upon the type of instrument used and the extent to which the records are dissected. Pluviographs are essential to describe adequately the precipitation in a storm and are located in conjunction with the standard gauges in strategic positions. As a working guide, ten per cent of the precipitation stations should be of the recording type in warm climates and five per cent in cold. [5] Provision may have to be made against freezing and against strong winds.

Storage precipitation gauges are of value in sparsely settled or remote areas, such as deserts or mountainous country. They are measured at weekly, monthly, or seasonal intervals. Light oil, chemicals, and antifreeze are used to prevent evaporation and freezing. Long term pluviographs are used in remote areas and give valuable details not otherwise obtainable. High costs limit the number of these which can be installed.

Radar is also used to determine the pattern of precipitation. [6] It shows the areal variation of rainfall, variations with time, and the movement of the storm cells. The method is in the experimental stage, is expensive, and requires a reference network for calibration.

10.4 The analysis of precipitation data. The variations of precipitation with time and space complicate its analysis and demand a diversity of techniques. A brief description of the more important and useful ones will be given here. The theoretical treatment, other methods and detailed analyses are dealt with in specialised books [7–10] and articles in periodicals.

The emphasis on precipitation in hydrology has arisen because records can be obtained easily and cheaply. In many regions they have been maintained for long periods and over extensive areas. In some countries they are the only available record. Precipitation depths can be converted logically to stream flows and hydrographs derived with an accuracy approaching that of some methods of stream flow measurement. By appropriate methods, precipitation can also be extrapolated rationally to an extreme or limiting value, the probable maximum precipitation, which has many uses in flood estimation. Statistical methods are readily applicable to precipitation data.

The average rainfall, either annual or monthly, is usually expressed as the arithmetic mean of a number of years or months. The length of record necessary to give a stable mean is related to the variability of the precipitation which in turn depends upon locality and many meteorological factors. Table 1.1 is a guide but usually 30 yr are considered a minimum period. Determination of the extremes or the nature of the statistical distribution of precipitation requires as long a record as possible. When information on the variation of rainfall with area is required say, by using average isohyetal maps, it is good practice to use a standard period of record, usually 30 yr, say 1931–60, for all stations. The particular period chosen can then be related, for key stations, to the long term averages or trends of the region and an assessment made as to how representative of the long term average was the 30 yr period chosen.

The median rainfall is often used instead of the arithmetic average, as it better represents the central tendency and is not affected to the same extent by extreme values.

A simple extension of this form of analysis gives the quartile range as a measure of variability. A frequency analysis of rainfall gives a full description and is a useful form of presenting precipitation data. [9]

10.5 The chronological analysis of precipitation. In hydrology, the ordering of precipitation with time is often required and can be done in a number of ways. The simplest is a bar chart in which the total precipitation during a period is plotted as a height against a width equal to the period in which the total was measured. Usually equal periods or bars are chosen. The annual rainfalls at a station over a series of years is shown as a bar chart in Fig. 10.3. Alternatively, vertical lines or the tops of the bars are drawn. Again, points representing the rainfalls may be joined by lines. Monthly rainfalls plotted as a bar chart show its seasonal incidence, and are used in planning and estimating the applications of water for irrigation or in determining the runoff from catchments.

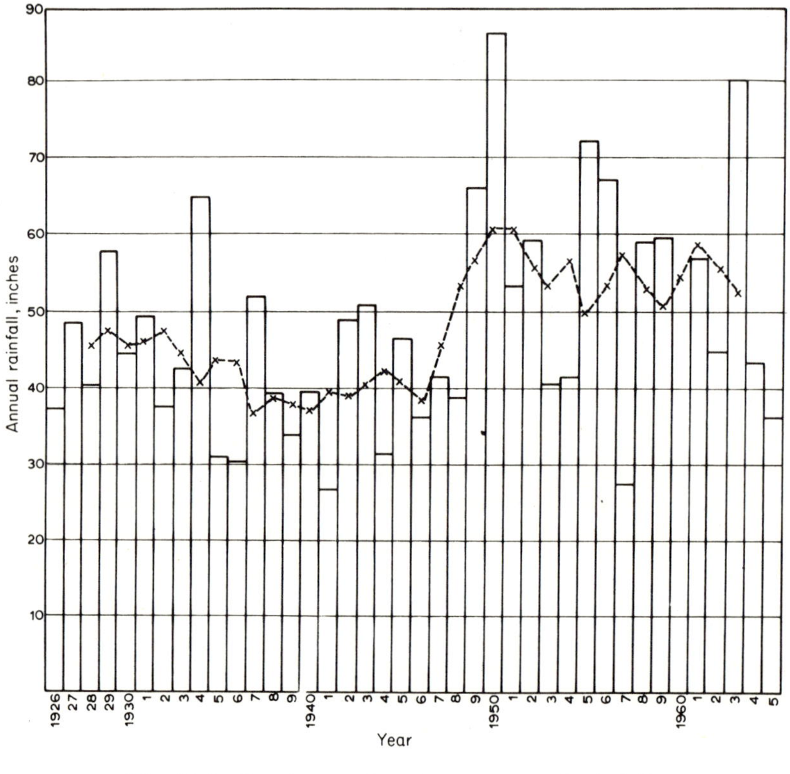

Fig. 10.3. Bar chart and moving average of rainfall.

Moving averages are used sometimes to damp out the fluctuations and extremes in an attempt to show trends. If the smoothing period chosen is n yr, then the n yr average

$$\frac{P_1 + P_2 + P_3 + \ldots + P_n}{n} \quad \text{or} \quad \frac{\sum_1^n P}{n}$$

is plotted at the $n + 1/2$ year and the next member is

$$\frac{P_2 + P_3 + P_4 + \ldots + P_{n+1}}{n} \quad \text{or} \quad \frac{\sum_{2}^{n+1} P}{n}$$

and is plotted at the $\frac{n+1}{2} + 1$ year, etc (10.1)

for example, a 5-yr moving average is

$$\frac{P_1 + P_2 + P_3 + P_4 + P_5}{5}$$

plotted at the third yr with the next member

$$\frac{P_2 + P_3 + P_4 + P_5 + P_6}{5}$$

plotted at the fourth yr. Figure 10.3 shows the 5-yr moving average plotted as a series of lines joining the 5-yr averages at adjacent intervals. It should be noted that, in moving averages, all years do not contribute equal weight, the first and last having the least effects. Also, very high or very low measurements affect all the averages into which they are incorporated and may by themselves, indicate a trend not otherwise apparent. However, trends occurring over a considerable part of the record are usually significant. Trends indicated by moving averages may be correlated with changes in other variables such as temperature, runoff, groundwater level, or vegetation. Moving averages also indicate the relationship between a specific period and the complete record at a station. Similarly, if the water potential of an area with short records is to be assessed, these records can be compared with the longer length climatic trends at nearby regions by using moving averages.

10.6 The time pattern of storms. On a smaller time scale, similar types of graphs may be constructed to show the temporal pattern of storms. Against the clock time, a graph of cumulated rainfall is plotted to give a mass curve, or the rainfall intensity plotted to give a hyetograph. Each can be converted to the other and both are shown in Fig. 10.4. The basic data is obtained from a series of rainfall observations or from a pluviograph, some types of which give mass curves directly.

In constructing a mass curve showing all possible detail, the cumulated rainfall is assessed and plotted at each time when there is a change in rainfall intensity. The smallest time interval which gives a constant slope, or uniform rainfall intensity is used and depends upon the sensitivity of the instrument used. It may be as small as 5 min. Quite often the nature of the investigation does not demand detail, and mass curves are constructed at wider intervals of, say, 1 hr, 3 hr, or 6 hr. In Fig. 10.4 a smoothed hyetograph, over hourly periods, is also shown. Wider intervals are sufficient for large areas and long duration storms.

In a mass curve, the graph ascends while it is raining with the slope representing the intensity in inches over the period. When the rain stops, the graph is horizontal. For a particular storm the mass curves at different locations give information on its nature, movement and development. A comparison of mass curves by plotting them on transparent paper and then superimposing show these features and often display a constant storm pattern. Complex storms and topographical influences confuse the pattern. Another method of comparing mass curves of the same storm expresses the total precipitation as 100% and graduates the vertical scale in proportion. This is shown in the right hand side of Fig. 10.4. The percentages of the total storm received in specified intervals may be compared and the maximum intensities for given durations determined. A simple way of finding the maximum intensity in a given duration uses a transparent scale marked with the required duration and finds the maximum vertical intercepts when this slides across the mass curve.

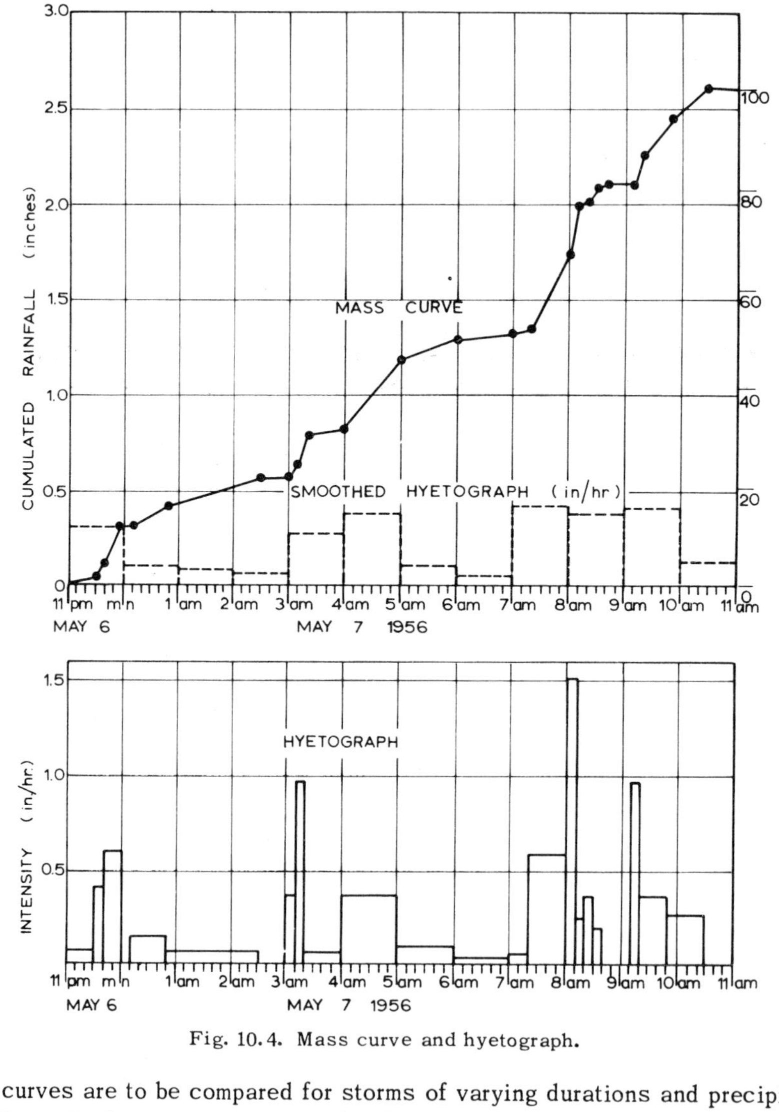

Fig. 10.4. Mass curve and hyetograph.

If mass curves are to be compared for storms of varying durations and precipitation depths, a dimensionless mass curve can be drawn in which the total precipitation is expressed as 100% and the total duration also as 100%. A study of about 80 short duration storms in the United States[11] gave the relationships

$$Y = 3{\cdot}25\,X^{0.831} \text{ from } X = 0 \text{ to } 38\% \qquad (10.2)$$

$$Y = 14{\cdot}90\,X^{0.413} \text{ from } X = 38 \text{ to } 100\% \qquad (10.3)$$

where Y is percentage of total storm precipitation, and X is percentage of total storm duration.

Where there are insufficient pluviographs to give detail for mass curves, a process of synthesis may be used.[12] The storm pattern is built up from available data such as distant pluviograph records, daily rain gauge readings and notations, synoptic reports, verbal or written information, etc. Isohyetal maps for the total storm or portions of it are helpful as are maps showing the start and finish of the more intense bursts, of rainfall. Synoptic charts assist the synthesis by indicating likely storm mechanisms and movements. General relationships such as equations (10.2) and (10.3) should only be used when no other data is available.

The time pattern of a storm is also shown by the hyetograph in which the rainfall intensity in in./hr is plotted against the time during which the intensity is current. It is derived from the mass curve and shown in Fig. 10.4. When the greatest possible sensitivity is required, the smallest interval over which the rainfall rate is constant is selected. Here, the mass curve appears as a straight line. This interval may be as small as 5 min, dependent on the sensitivity of the instrument used, and is varied, as shown in Fig. 10.4, to give the fullest possible detail. It is often more convenient, however, to use a uniform interval suited to the project and type of storm. This smooths the pattern as shown in the dotted lines in Fig. 10.4. In many hydrological problems such as the estimation of runoff or the unit hydrograph, the minor fluctuations in rainfall intensity are not significant. The distribution of rainfall over the catchment and the processes of interception, infiltration, sub-surface flow, interflow, and ground water movement all tend to damp out the short period variations in rainfall. Small area storms are frequently analysed at one-hr intervals and, over larger areas, 6-hr intervals are often sufficient. Short duration rainfall intensities are important in urban drainage with its problems of gutter flow and culvert design. Thus in urban applications detailed rainfall analysis is usually required.

The hyetograph has the advantage that it uses a larger vertical scale than the mass curve, and is appropriately used and compared with the hydrograph of stream flow. It clearly shows the variations in rainfall intensity and is adaptable to computer programmes as in the determination of unitgraphs.[13] The area under the hyetograph represents the total precipitation received in the period.

Residual mass curves, or mass curves in which the cumulated departures from the mean are plotted against the time, are a compact form of showing the variation of precipitation with time, and permit the use of a larger vertical scale. They are usually used with monthly or annual figures rather than to show variations within storms. The vertical scale can be expressed as a percentage of the mean to give better comparisons of variability between places of different average rainfalls. The amplitude of the residual mass curve indicates the extent of the deficiency below, or surplus above, the mean. A small amplitude indicates low variability of rainfall. An indication of trends is given by the residual mass curve. For example, a steadily rising curve indicates a continuation of above average rainfall or a wet spell, a falling curve shows a series of dry years, and a straight line, average conditions. Residual mass curves of rainfall are useful in studies of drought. The residual mass curve of streamflow is used to determine the yield of a stream. The relationship between two residual mass curves may be used to extrapolate the smaller length record.

10.7 The average temporal pattern of storms. In obtaining estimates of surface runoff from rainfall data, the time pattern of the storm, as given by the mass curve or hyetograph, is significant. Statistical analyses of rainfall data give the total amount of rainfall for a given duration and for a certain frequency of occurrence or return period, but not the distribution of rainfall in time. That is, although the once in 20 yr, 6-hr rainfall is estimated as 12 in., the spread of this 12 in. during the 6 hr is not known but required.

A direct approach examines the mass curves of all storms of the given duration and similar intensity, to determine whether they possess a typical temporal pattern which can be accepted as the design pattern. Where the data are available, this is the best approach.

A mass curve may also be synthesised from synoptic reports and other data as explained earlier. [12] Some types of storms possess characteristic patterns. For example, the frontal cyclone is marked by light rain which increases with the passage of the warm front, and after a period of intermittent rain, is followed by the heavy precipitation of the cold front and the showers of the cold streams. This can be expressed as a synthetic mass curve and applied to this type of storm.

The average temporal patterns of storms may also be used. These have been determined for two principal types, the thunderstorm or convergent storm model of 1 hr duration [4] and the more general storm of 6–24 hr duration. [14] Dimensionless mass curves of these two types give the percentages of the total storm depth in terms of the percentages of the total storm duration. A large number of United States storms give the average patterns shown in Fig. 10.5. Thunderstorms of 1 hr duration show a marked peak at the beginning of the storm, and as intensity increases from 1 in./hr to 5 in./hr, this becomes less marked. General storms of 6, 12, 18 and 24-hr duration show an average relationship, irrespective of the duration, with the most intense rainfall falling during the third quartile. [14] Although these curves conceal many variations in the time distribution of rainfall, they suggest an average pattern which might be used when other information is not to hand. Storms of 1 hr and from 6–24 hr duration follow the patterns indicated. For storms between 1 and 6 hr duration, the time pattern should conform to the more probable type, either thunderstorm or general storm. If this is not known, the peak intensity should be considered as moving with increased duration from the first to the third quartile and as becoming more uniform when the rainfall intensity increases. An appropriate mass curve may be estimated in this way from Fig. 10.5. Storms exceeding 24 hr in duration are usually complex, involving more than one storm mechanism but if these are separated out, a likely pattern can be deduced. Otherwise actual storms must be relied upon for a pattern.

Topography of a basin and its position relative to the flow of moist air and the storm systems affect the temporal pattern. In such cases, and as a general rule, the common temporal patterns of the storms within the basin are preferred.

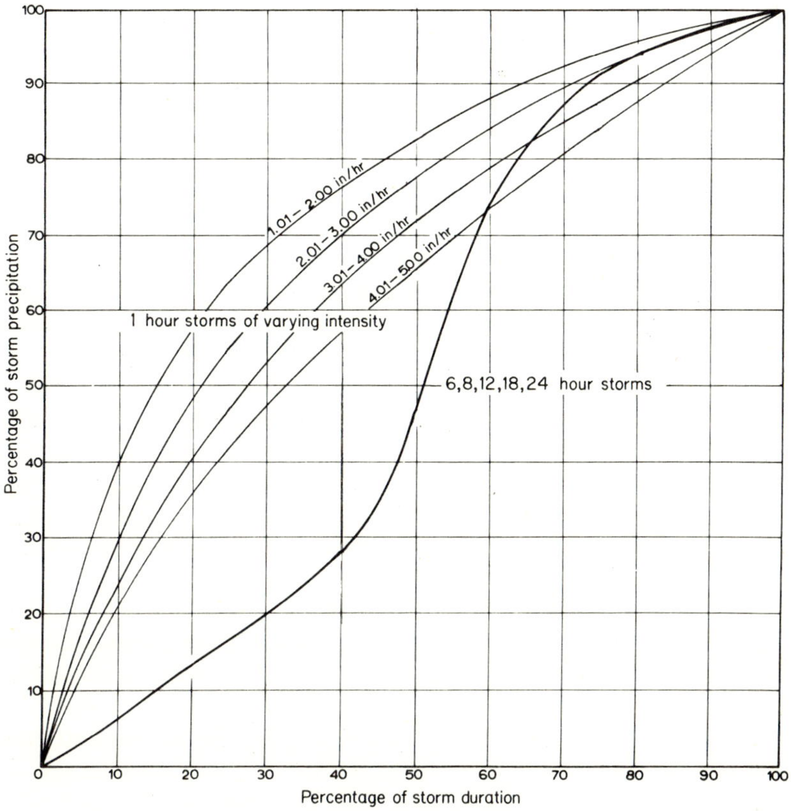

Fig. 10.5. The temporal patterns of storms.

10.8 The 'water year'. Precipitation and runoff are related and, in many regions, both have a marked seasonal distribution. In compiling and analysing yearly rainfall and runoff records, it is an advantage to select a twelve monthly period such that the runoff is caused by the rainfall within those months. That is, the period encompasses the wet season and does not divide it. This is termed the 'water year'. It commences in a month when stream flow is normally at a minimum and before the seasonal rains have begun. For example, in Northern Australia, the rains commence in November or December, and continue until May. A calendar year, January to December, would include parts of two wet seasons. Here an appropriate 'water year' is the twelve months ending 31st. August.

Runoff usually increases some time after the start of the wet season because of the necessity to replenish soil moisture but continues into the dry season. However, successive wet seasons are nearly always sufficiently wide apart so that the one defined 'water year' encompasses both the seasonal runoff and rainfall.

When rainfall and runoff are distributed uniformly over the year, or if no marked wet season occurs, there is no clearly defined water year, and data is collected in the calendar year.

10.9 Frequency analysis of precipitation data. As well as the order in which individual bursts of rainfall occur or the storm's temporal pattern, these bursts may be considered from the point of view of their frequency of occurrence. That is, the intense rainfall bursts for a given duration are statistically analysed to give the depths which may be reached or exceeded, (a), in a certain percentage of years, (b), as a probability of occurrence, or (c) in a number of years, termed the return period or recurrence interval. For example, a heavy fall of 1 in. in 30 min, intensity 2 in./hr, may occur in the third quarter of a 6-hr storm, and is of interest if the chronological pattern is required. However, if the interest is in how often such a fall may occur, all heavy 30 min duration bursts within the length of record at the station are considered statistically to find, perhaps, that a 30 min duration fall of intensity of at least 2 in./hr can be expected to occur in 20% of the years, with a probability of 1 in 5, or once in every 5 yr. This means that in 50 yr of records, about 10 similar or higher falls have occurred and, in the future 50 yr, as many again might be expected. The return period or recurrence interval, is 5 yr, and the once-in-5-yr 30 min rainfall is at the rate of 2 in./hr. Provisos and modifications to these simple statements can be made in the light of modern statistical theory which is covered in appropriate texts. [7–10]

Information of this type on the frequency of occurrence of rainfall of certain durations and intensities is required in the design of drainage works, including gutters, sewers, waterways, channels and culverts, or in the estimation of floods which might be expected by levees, spillways, terraces, and conservation works. The return period selected is often decided by balancing benefits against costs, but economics, local policy, and other factors may play a part in the decision reached.

The duration of the rainfall burst used is related to the time of concentration of the catchment and so, approximately to the size of the basin. Short duration rainfalls of 5 min to 2 hr apply to small areas, while those up to 72 hr are used in large catchments and major flood studies.

Two methods of frequency analysis are used here, that of placing the rainfall bursts into class intervals, and the second of ranking them in order of intensity. Both methods are related and give similar results although the ranking method is often more convenient to use. Details and discussions of methods, theories and the fitting of data to the various types of statistical distributions are covered in specialised texts. [7–10] Simple applications to rainfall data are considered here.

10.10 Classifying rainfall data into class intervals. The standard statistical method of obtaining a frequency distribution by placing members of a series into class intervals is often used for rainfall data. [8] Rainfall records of the required duration say annual, monthly, daily, 5 min etc, are placed into class intervals. Usually about ten classes are satisfactory. For example, the maximum 5 min rainbursts occurring in each year at Sydney from 1913–1962 can be placed in class intervals as in Table 10.2. The data is cumulated to give the number greater or less than the lower bound of the class, and then plotted as this number against the class interval on graph paper. A straight line or a curve smooths the data and gives a better description of the distribution. Alternatively, the various moments of the data may be determined and used to define the probable theoretical distribution, the properties of which may then be used.

Table 10.2. *Frequency of the annual maximum 5-min rainbursts, Sydney,* 1913–1962

Class interval (0·01 in.)	*Frequency* (no. of yr)	*Cumulated yr more than*	*Cumulated yr less than*	*% yr more than*	*% yr less than*
16–20	1	50	0	100	0
21–25	5	49	1	98	2
26–30	9	44	6	88	12
31–35	8	35	15	70	30
36–40	12	27	23	54	46
41–45	6	15	35	30	70
46–50	5	9	41	18	82
51–55	1	4	46	8	92
56–60	0	3	47	6	94
61–65	3	3	47	6	94
		0	50	0	100

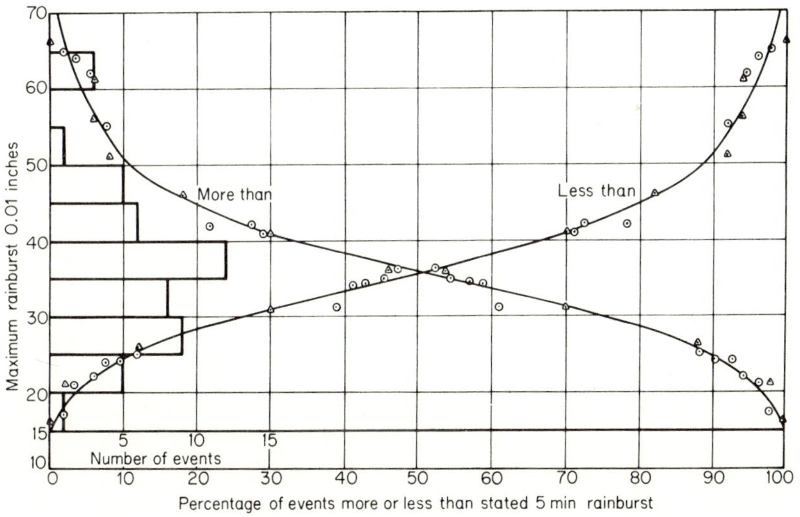

Fig. 10.6. Frequency curves.

Figure 10.6 shows the data of Table 10.2 plotted on linear graph paper. The distribution appears approximated by a skewed curve. A plot on Normal probability paper in Fig. 10.7 shows that the data does not fit well to a straight line. By plotting the logarithm of the interval against the probability scale, also in Fig. 10.7, it is seen that a straight line is a reasonable fit and thus the data is approximately fitted by a log Normal distribution. The square root or any other function of the rainfall may also be plotted against the Normal probability scale to test whether a better fit can be obtained. In a similar way, graph papers with other distributions represented along one axis so as to give a straight line plot are also convenient to use. One employing the reduced variant $(X - \bar{X})$ or a

Fisher Tippett Type 1 or Gumbel distribution has been found convenient in hydrology and is shown in Fig. 10.8. The fitting of the data to the theoretical distributions can also be achieved by statistical methods using the moments of the distribution. Goodness of fit can also be determined by appropriate tests. However, the use of the various types of probability paper and fitting the data by a straight line is a quick method of testing and smoothing the data so that it more nearly approaches the population. Its limitations should be realised and extrapolation outside the range of the data is doubtful and is not recommended.

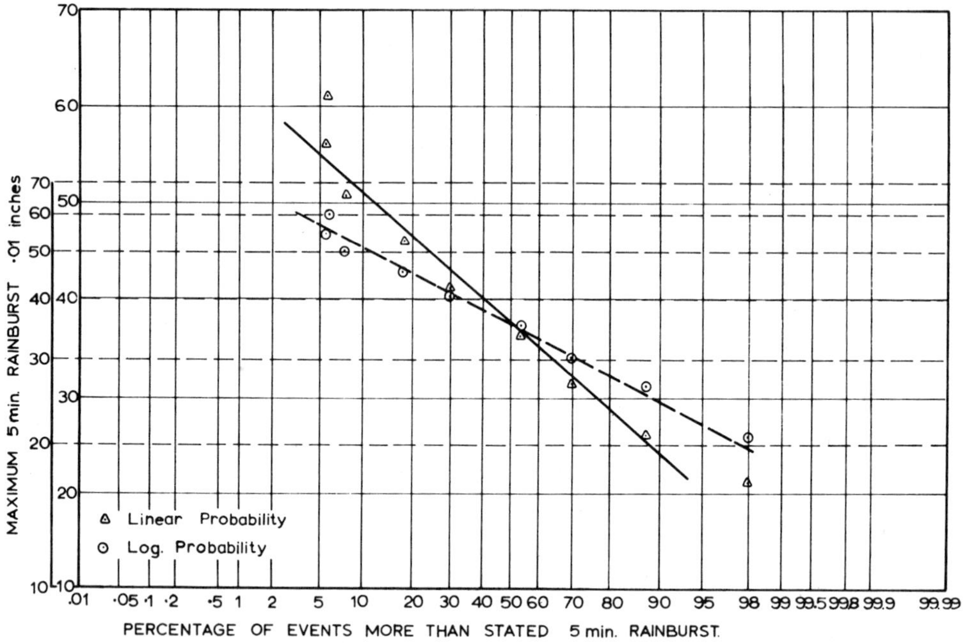

Fig. 10.7. Normal probability paper—linear and log plot.

The distributions applicable to a set of data may vary with space and time. For example, annual rainfall data in Australia has been found to be Normally distributed in some locations, but not in others. Also, Australian monthly rainfalls can be converted to Normal by a logarithmic transformation and the daily rainfalls fitted by the Gamma distribution. [15]

The method of class intervals is suited to the full series of events and the curves plotted are frequently referred to as duration curves. These have applications in surface water hydrology in determining the available water of a stream for irrigation, power, or town supply. With the use of the full series of data, there is sometimes doubt as to whether the events are independent. Extreme values, on the other hand, for example the annual maximum 5-min rainfalls, are independent of the other maximum events in the adjacent years.

10.11 Ranking rainfall data. Rainfall data may be ranked in order of magnitude and a statistical distribution fitted to the ranked data. A rank or order number, m, is assigned to each member of an array or ordered series of data covering n yr. The return period, or recurrence interval, T_r, is defined as the average number of years within which a given event is equalled or exceeded. Each member of the ordered series is assigned a return period which is based on its rank number, m, and the number of years in which it has been ranked, n. Various formulae are proposed but

$$T_r = \frac{n+1}{\mathrm{m}} \qquad (10.4)$$

has the least objections. The tenth member of a 25-yr series thus has a return period of $25 + 1/10 = 2{\cdot}6$ yr. Instead of the return period, the frequency of the event or the probability of it not being reached or exceeded may be used. This is given as the reciprocal of the return period. Expressed as a percentage,

$$\text{probability of the event or less} = \frac{100\mathrm{m}}{n+1} = \frac{100}{T_r}. \qquad (10.5)$$

It should be noted that the probability of the event being reached or exceeded or P excess equals one minus the probability of it not being reached, or P less. Expressed in percentages,

$$\%\,P \text{ excess} + \%\,P \text{ less} = 100. \qquad (10.6)$$

The ranking method gives essentially the same result as the class interval method. If the data conforms to one of the statistical distributions mentioned above, that is, Normal, log Normal, or reduced variate, statistical techniques may be used, or probability paper, to fit the data to the chosen distribution.

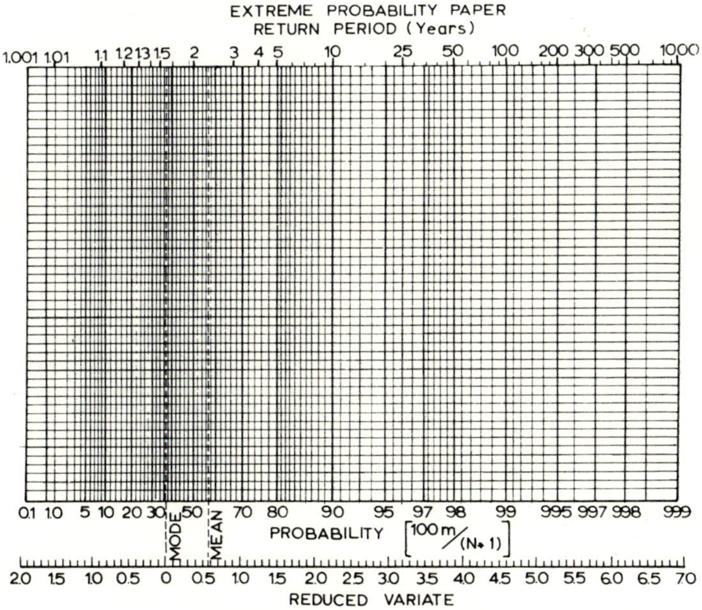

Fig. 10.8. Reduced variate probability paper.

As probability and return periods are reciprocals, different scales are required on the special probability papers. This represents no problem and one can be converted to the other. Both scales are shown in Fig. 10.8.

A selection of the data used in Table 10.2 is ranked and return periods and percentage probabilities computed in Table 10.3.

Table 10.3. *Ranks, return periods, and probabilities of events more than or less than selected annual maximum 5-min rainbursts*

Year	Magnitude	Rank	Return period $T_r = \frac{n+1}{m}$	Probability more than P excess $= \frac{100m}{n+1}$	Probability less than P less $= \left(1 - \frac{m}{n+1}\right)100$
1926	65	1	51	2·0	98
1922	64	2	25·5	3·9	96·1
1914	62	3	17	5·9	94·1
1938	55	4	12·8	7·8	92·2
1917	42	11	4·6–3·6	21·6–27·5	78·4–72·5
1919	42	11	4·6–3·6	21·6–27·5	78·4–72·5
1949	42	11	4·6–3·6	21·6–27·5	78·4–72·5
1951	42	11	4·6–3·6	21·6–27·5	78·4–72·5
1959	41	15	3·4	29·0	71·0
1954	36	27	1·9	52·9	47·1
1934	35	28	1·8	54·9	45·1
1944	34	29	1·75–1·7	56·9–58·8	43·1–41·2
1958	34	29	1·75–1·7	56·9–58·8	43·1–41·2
1928	33	31	1·6	60·8	39·2
1918	25	45	1·13	88·2	11·8
1916	24	46	1·11–1·08	90·2–92·2	9·8–7·8
1941	24	46	1·11–1·08	90·2–92·2	9·8–7·8
1920	22	48	1·06	94·1	5·9
1935	21	49	1·04	96·1	3·9
1946	17	50	1·02	98·0	2·0

The information of Table 10.3 plotted on Fig. 10.6 gives a similar fit to the lines established by the class interval method as might be expected. The ranking method gives a larger number of points, is easily computed and plotted but involves additional work in ranking the data. In using the method of ranking, all members or a selection of the series may be plotted. Also, historical data can be incorporated in a series, even if it is not continuous with the other data, provided that its magnitude, order, and return period is known. For example, if the highest known rainfall or flood height occurred 200 yr ago, and has not been exceeded since, it can be assigned the return period of 1 in 201 yr, even although continuous records have only been made for the last 100 yr. This information may be plotted with other data to extend the length of record.

10.12 Basic stage and annual series. Data on extreme events can be assembled in two ways. If the highest in each year is selected, an annual series is obtained. This may exclude lower events in some years which exceed the annual events of other years. By considering the lowest recorded annual event a series may be set up incorporating all events above this or another selected basic stage. This is called a basic stage series, or a partial duration series. It is necessary to assure that consecutive events are independent before including both in the series.

Both annual and basic stage series are analysed statistically, but give different results in the lower return periods although they converge beyond the 10-yr return period. Theoretical considerations give the relationship between the return periods of the two series in Table 10.4.

Table 10.4. *Corresponding return periods in years*

Basic stage series	0·5	1·00	1·45	2·0	5·0	10·0	50·0	100·0
Annual series	1·16	1·58	2·00	2·54	5·52	10·5	50·5	100·5

The type of series used depends upon the data available, the purpose of the investigation, and on personal choice. For extreme events, exceeding 10-yr recurrence interval, the annual series is usually preferred as it is easier to compile. For more frequent events the basic stage method gives a more accurate assessment of the requirement. It involves considerably more work than the annual series, although it is possible to convert the annual return period to its corresponding basic stage return period.

Empirical factors have been determined in the United States [18] to enable the conversion of the values of an annual series to those of a basic stage series and vice versa. They are given in Table 10.5.

Table 10.5. *Factors for converting annual and basic stage series*

Return period	*Annual to basic*	*Basic to annual*
1 in 2 yr	1·14	0·88
1 in 5 yr	1·04	0·96
1 in 10 yr	1·01	0·99

For example, if the once in 2, 5 and 10-yr annual values of 60 min rainfall are 2·5, 3·0, 3·5 in./hr the basic stage once in 2, 5 and 10-yr values of 60 min rainfall are $2{\cdot}5 \times 1{\cdot}14 = 2{\cdot}75$; $3{\cdot}0 \times 1{\cdot}04 = 3{\cdot}12$ and $3{\cdot}5 \times 1{\cdot}01 = 3{\cdot}54$.

10.13 The frequency of bursts of rainfall of a given duration. The mass curves of heavy storms show bursts of intense rainfall over various durations which may be from 5 min to the length of the storm. The heaviest burst over the given duration is the maximum for that storm. From all the storms in a year the annual maximum bursts for that duration or all the independent bursts above a basic stage may be obtained. If all years of record are considered an annual series or a basic stage series is compiled. These series may be analysed by the methods already described to obtain the frequency of occurrence of rainfalls greater or equal to a given rain burst of stated duration. Alternatively, it is possible to derive the rainfall intensity for the stated duration which can be expected in a given return period. That is, an analysis of the 15-min rain bursts may show that falls of at least 4 in./hr over 15-min periods occur once in 5 yr or alternatively, that once every 5 yr a rainfall intensity of at least 4 in./hr may be expected over a 15-min period.

Where pluviographs have been maintained for a sufficient number of years, they may be dissected to give maximum rainbursts over the separate durations. Analysis of these gives the frequency of the rainbursts of various intensities for these durations. If the areal cover of the pluviographs is adequate, the information is most suitably expressed in the form of isohyetal maps which may then give the rainfall intensity for given durations and stated return periods. [18] For example, a country wide analysis of pluviographs enables the plotting, on a map, of the once-in-10-yr, 30-min rain burst. Isohyets may be drawn and, if the rainfall intensity for once-in-10-yr 30-min duration bursts are required at any place on the map, interpolation between the isohyets is satisfactory, except in mountainous regions.

The intense rainfall bursts occur over a wide range of durations and each of these has an appropriate frequency analysis with associated return periods. There are many combinations of rainfall durations and return periods and a multiplicity of maps would result if all were analysed in the above way. To minimise computation and maintain consistency, theoretical and empirical methods have been used to establish the relationships between rainfall intensity, rainburst duration, and return periods. Furthermore, where pluviograph records are limited in areal coverage and length of record, use may be made of generalised relationships to interpolate between the known factors and save considerable work. For example, if the once-in-2-yr and once-in-100-yr rainfall bursts are known, the bursts for other recurrence intervals could be deduced from an appropriate relationship between intensities and return periods. Also, if the once-in-2-yr, 1-hr and 24-hr rainbursts are known, it is possible to deduce the rainbursts for other durations from a knowledge of the relationship between rainburst intensity and duration.

For simplicity, the relationships of rainburst intensity to duration and of rainburst intensity to frequency of occurrence or return period are considered separately.

10.14 Relationship between rainburst intensity and duration. Analysed rainbursts show that the shorter the duration, the greater the rainfall intensity. If the rainbursts in a storm or the maximum rainbursts for the various durations, e.g. 5 min, 10 min, ----, 24-hr etc., are plotted against the total depth, P in., or against intensity, i in./hr, a depth-duration or intensity-duration graph, similar to Fig. 10.9 is obtained. The inverse variation between rainfall intensity and duration enables formulae to be established and the constants determined from pluviograph data for particular locations.

The general form of the equation is

$$i = \frac{c}{(D + b)^n} \tag{10.7}$$

where c, b, and n are constants and D duration in min.

Although values of n and b appear reasonably constant for all return periods, different values of c are needed. Figure 10.11 shows that the less frequent storms may be represented by almost parallel curves placed higher in the axes suggesting that formulae involving the recurrence interval be used,

viz:

$$i = \frac{cT_r}{(D + b)^n}\,. \tag{10.8}$$

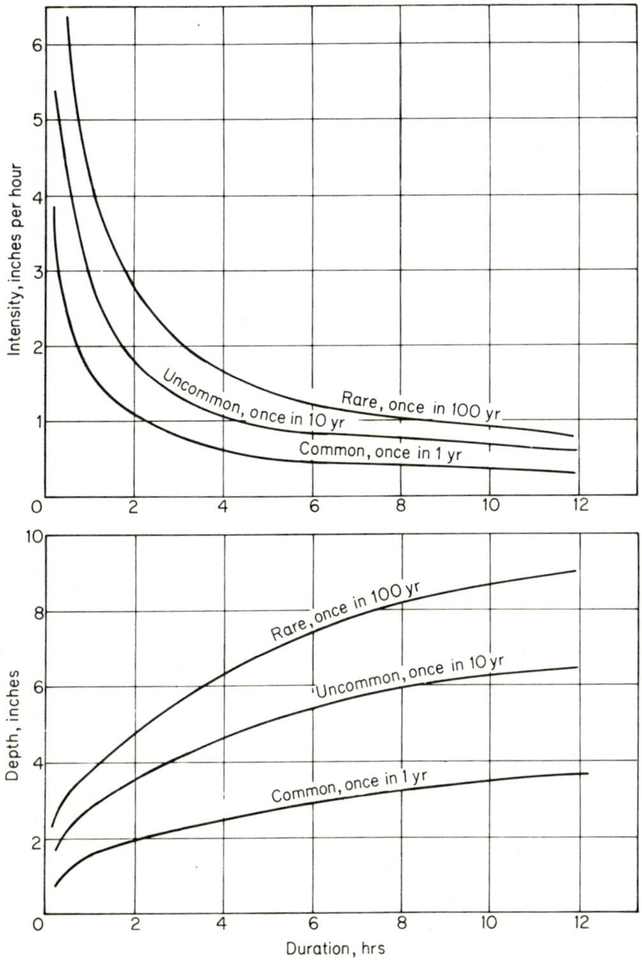

Fig. 10.9. Rain burst depth against duration, rain burst intensity against duration.

In Australia at capital cities,[16] for the once-in-one-yr rainburst, c varied from 3·88 to 31·70, b from 0 to 8, and n from 0·523 to 0·760. A value of $b = 5$ was accepted for all places, while c and n were determined from the once-in-one-yr 5-min and 3-day rain bursts at a network of stations. Isopleths of c and n were constructed, so that c and n could be estimated at all localities.

Rainfall intensities for the once-in-one-yr rainburst for other durations is then deduced from equation (10.7). In the U.S. and British Isles, values of c, b and n of a similar order were determined. For Boston it was found[17] that $c = 16\,T_r^{0\cdot27}$, or $c = 16$ for the once-in-one-yr flood, $b = 7$, and $n = 0\cdot7$.

This gives for Boston,

$$i = \frac{16\,T_r^{0\cdot27}}{(D + 7)^{0\cdot7}} \; . \tag{10.9}$$

Using maximum values of rainfall intensities the exponent 'n' approximates to 0·5. This is discussed further in paragraph 18.6. Thus, the relationship between intensity and duration is established from the formula

$$i = \frac{c}{(D + 5)^n}$$

by determining c and n from the data for two known durations and return periods at the site in question. [16]

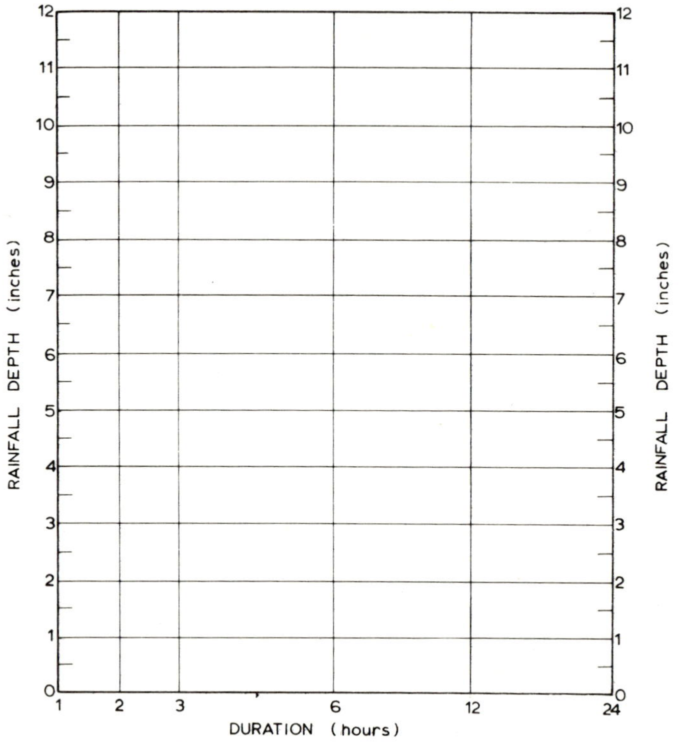

Fig. 10. 10. Relationship between rain burst intensity and duration of burst.

Instead of a formula, an empirical depth-duration scale has been devised so that the depth of rainfall for various durations plots as a straight line. [18] Figure 10. 10 shows this scale for some 200 U.S. first order stations. It is claimed that this empirical relationship holds for that continent for the same return period. From a knowledge of the depth for two durations, say the 1-hr and 24-hr once-in-2-yr rain burst, the depth for other durations and once-in-2-yr may be read off from a straight edge placed on the two known points. The rain bursts for durations of 30, 15, 10 and 5 min are obtained by multiplying the 1-hr rain burst by the ratios 0·79, 0·57, 0·45 and 0·29. Although the relationship shown by this scale applies strictly to the United States, it can be used in other countries. The small errors can be allowed for by slight changes in the duration scale. [20]

Where pluviographs are available, the maximum rain bursts of a given return period for two durations, say 24-hr and 1-hr, are easily analysed and the scale of Fig. 10.10 used to obtain bursts for other durations. If daily raingauge readings only are to hand, the 24-hr once-in-2-yr burst can be determined and the one-hr, once-in-2-yr rainfall obtained from an empirical relationship between the mean of the annual series of maximum daily precipitation amounts, and the mean annual number of thunderstorm days. This is shown as Fig. 10.11. [20]

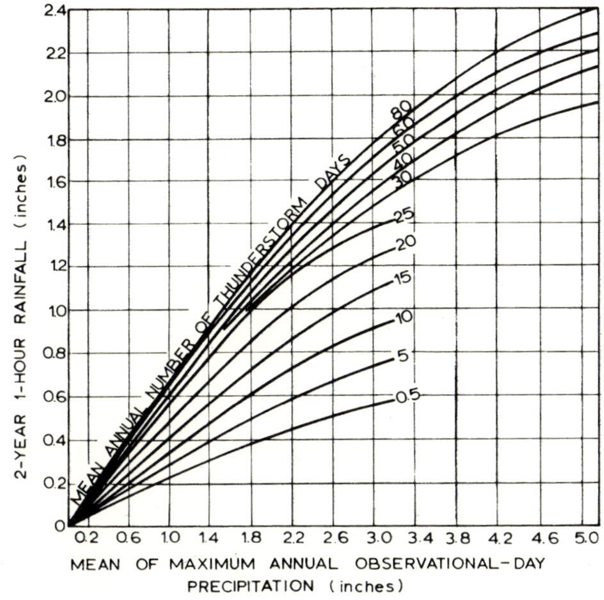

Fig. 10.11. Empirical relationship for 2 yr/1 hr rainfall.

Another empirical relationship, established in Australia, [16] gives the once-in-one-yr 5-min rainfall intensity as

$$i = \frac{0 \cdot 25\, \bar{P}^{n}}{d^{m}}, \tag{10.10}$$

where $\bar{P}$ = average annual rainfall in in. and d = average number of rain days greater than 0·10 in.

When $\bar{P} \leqslant 50$ in., $n = 0 \cdot 7$ and $m = \frac{4}{3} \cos^3 \phi$ (10.10a)

when $\bar{P} > 50$ in., $n = 0 \cdot 25$ and $m = \frac{31}{4} \cos^3 \phi$ (10.10b)

where ϕ is latitude in degrees. It must be remembered that the various formulae and diagrams described were determined empirically in certain locations and apply strictly to them only. It is better practice to use local data to verify and check the relationships and constants.

10.15 Converting clock hour and rain day falls into 60-min and 24-hr amounts. The majority of rainfall records are maintained on a clock hour or an observational day basis. Various analyses have been made to relate the amounts recorded in these periods to the maximum rainfall recorded in a 60-min or 24-hr period. It has been found[18] that an average correction factor of 1·13 applies to both durations. The clock hour or observational day rainfall data multiplied by 1·13 thus gives a better estimate of the maximum 60-min and 24-hr rain bursts.

10.16 Relationship between rain burst intensity and its return period. In paragraph 10.13 it was stated that where data was available, usually from pluviographs, the annual or basic stage series of maximum rain bursts of a given duration can be statistically analysed to give the intensity of the rain bursts in terms of return periods. However, if an empirical or theoretical relationship were established, it would give consistency to the data and enable estimates to be made from less complete information.

The relationship between rain burst intensity, duration and frequency, or return period, can be shown by plotting for all years of records, the rain burst intensities against their durations. If all high intensity storms are plotted in a diagram like Fig. 10.9 there will be a large number of rain bursts at the bottom of the graph with the rare ones at the top. If the series of rain bursts are analysed to obtain the frequency or return period of the rain bursts of each duration, the common ones may have a return period of say once in one yr while the rare ones perhaps one in 100 yr. This is shown in Fig. 10.9, where the linear scales indicate a series of hyperbolae, one for each return period. A log–log plot would give approximately parallel straight lines. As discussed before,[17] empirical formulae of the type

$$i = \frac{c\,T_r^{\,m}}{(D + b)^n}$$

have been determined with constants obtained from local analyses.

The theoretical frequency distributions of the maximum rain bursts have been examined and either the log–Normal or reduced variate, (Fisher Tippett Type 1) distributions have been found to be a satisfactory fit. These two distributions may therefore be used to interpolate or extrapolate the data to give other return periods when some return periods are known, or alternatively, when the moments of the distribution have been determined. Special probability papers enable the data to be plotted as straight lines and simplify interpolations and calculations. The moments of the distributions can be substituted in the appropriate equations if desired but considerable labour is involved. The preparation of various nomograms and diagrams have simplified the use of the two distributions. The relationships between the elements of a log–Normal distribution have been expressed in graphical form by McIlwraith[16] and Weiss.[19] In Fig. 10.12, for the reduced variate distribution, the relationships between the return period, T_r; length of record, n; mean, $\bar{X}$; standard deviation σ; and extreme value, X; are shown. If four of these are known, the other may be estimated. For example, if the mean of the annual maximum 24-hr rain bursts X is 5·4 in., their standard deviation 1·2 in., and there are 40 yr of record, then the once-in-10-yr rain burst, X, has a reduced variate, $X - \bar{X} = 2{\cdot}1$. That is, the once-in-10-yr 24-hr rain burst is $5{\cdot}4 + 2{\cdot}1 = 7{\cdot}5$ in. This nomogram also gives the

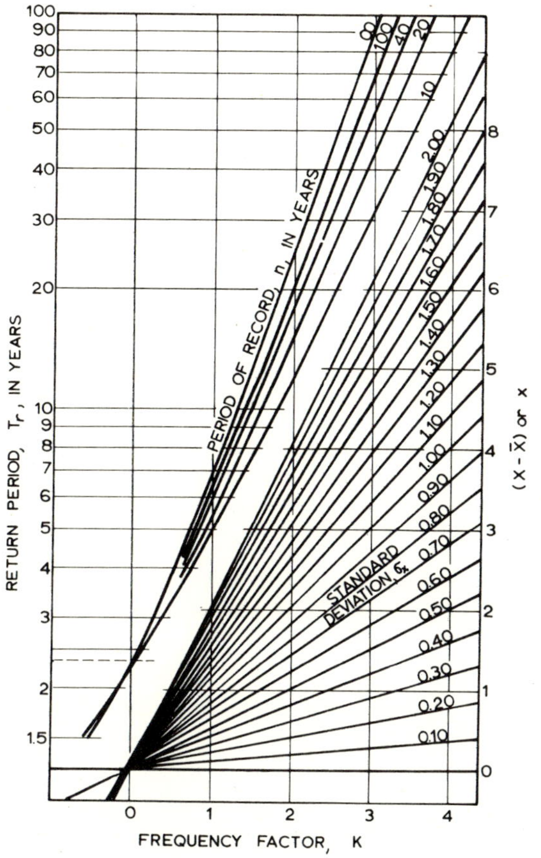

Fig. 10.12. Relationships between elements in a reduced variate distribution.

frequency factor K defined by $X = \bar{X} + K\sigma$ and used later in paragraph 18.8.

Although the log–Normal and reduced variate distributions are reasonable fits to the rainburst series, extensive analyses of United States storms have produced an improved scale, shown as Fig. 10.13, in which the rain bursts depths of given duration plot as a straight line against the various return periods. The spacing of the vertical lines, return periods, was based on the Fisher Tippett Type 1 distribution beyond 20 yr, but from 1 to 10 yr it was based on freehand curves drawn through plottings of partial duration series data. If the rain burst depth is known for two return periods, e.g. the once-in-2-yr and the once-in-100-yr, then the depths for other return periods can be read off from the straight line joining these two points on the nomogram.

Figure 10.13 strictly applies to U.S. data only, but has been used in Africa[20] with slight adjustments. In the absence of other information, the diagram gives a useful interpolation of the return periods. Rain burst intensities derived from Fig. 10.13 plot on log–Normal and reduced variate probability papers as approximately straight lines, showing that there are not large differences between the suggested methods.

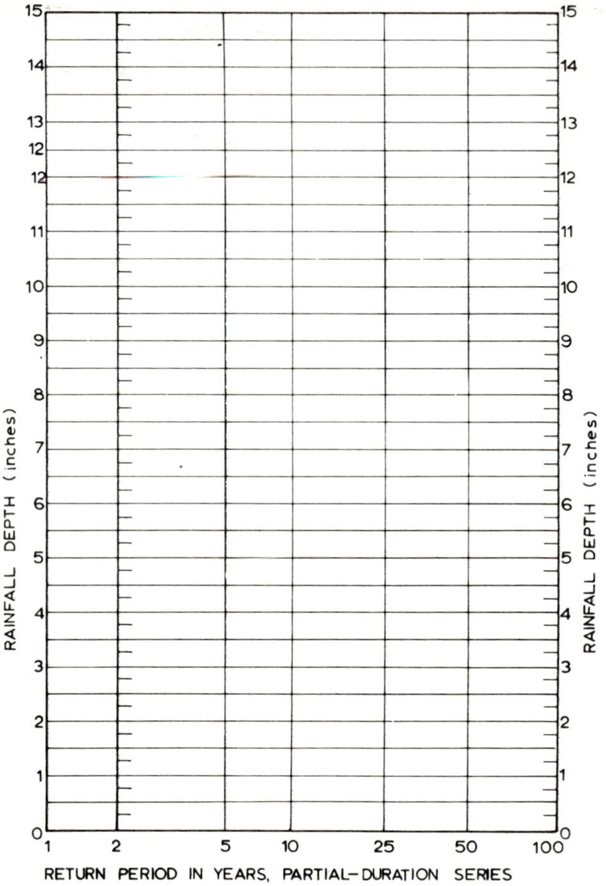

Fig. 10.13. Relationship between rain burst depth and return period.

Another mainly empirical relationship between the return period of a rain burst and the ratio of its intensity, or depth to the once-in-one-yr intensity or depth is given in Fig. 10.14.[21] In this, allowance has been made for places of high, moderate and low annual rainfall. From this diagram the once-in-100-yr rain burst is 2·2, 3·0 or 4·4 times the once-in-one-yr rain burst for areas of high, moderate or low annual rainfall. If the once-in-one-yr rain burst at a location is known and its average annual rainfall, the other return periods may be deduced. A generalised relationship of this type, although only approximate, is valuable in giving preliminary estimates.

10.17 The probability of occurrence of rain bursts in seasons. The maximum annual rain bursts, or those above a basic stage, are analysed to give their frequency of occurrence or return periods. These rain bursts also have a seasonal distribution, that is, more occur in some months than in others. In fact, some months may be completely free of all intense rain bursts. An examination of the members of the annual or basic stage series with respect to their times of occurrence enables the probability of their seasonal

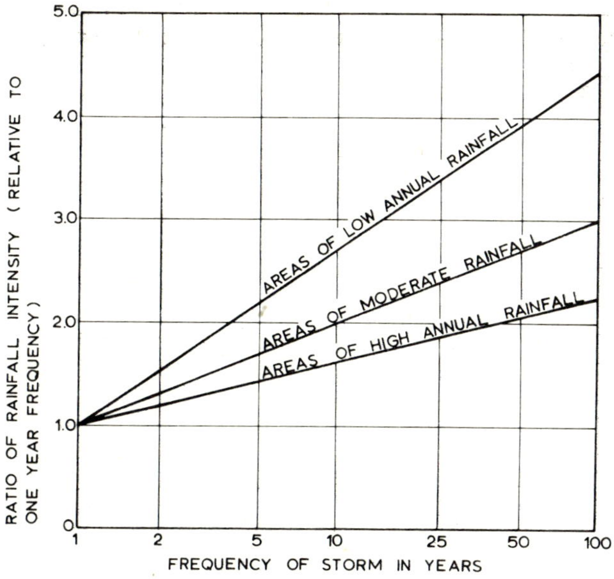

Fig. 10.14. Relationship between rain burst ratios and return period.

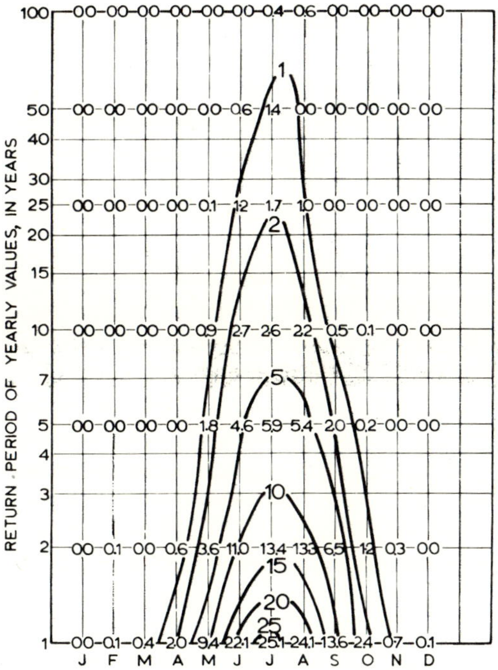

Fig. 10.15. Seasonal probability of intense rain bursts.

disposition to be decided. For example, if a series covering 100 yr of records, or a series extrapolated to that length, has the rain bursts in excess of the once-in-one-yr amount divided into their months of occurrence, the number of these in each month may be plotted on a horizontal time scale against a vertical logarithmic scale of the return period, shown in Fig. 10.15. Each figure along the once-in-one-yr return period line represents the percentage probability of rain bursts in excess of the once-in-one-yr amount occurring in that month.

The rain bursts in excess of the amounts necessary for other return periods, e.g. 5, 10, 25 yr may also be allotted to their months of occurrence and so the seasonal probability resolved. When the series are plotted for all return periods, isopleths of equal probability may be drawn.

A diagram similar to Fig. 10.15 can be deduced for each duration from the data of each meteorologically homogeneous zone. There is a different seasonal incidence for rain bursts of each duration. For example, the one-hr rain burst usually comes from a type of storm different from the 12 or 24-hr rain burst, and occurs typically in different months. [18]

Geographical variations in seasonal probability occur because the types of storm giving the significant rain bursts depend upon location and, especially, latitude. Each type has a particular seasonal incidence in an area. For example, tropical cyclones are restricted to the summer seasons and extratropical cyclones may only visit some latitudes with intensity in the winter.

From figures derived like 10.15, the monthly probability of the occurrence of the once-in-n-yr rain burst of that duration can be deduced by reading from the once-in-n-yr recurrence interval and interpolating for the respective months between the isopleths of probability. For example, the probability of the once-in-5-yr rain burst occurring in June, July and August are 5, 7, and 6% respectively.

Seasonal information on rain bursts is needed when flood flows are estimated from precipitation depths and infiltration indices. A higher rain burst in summer may, by virtue of greater infiltration, give a lower flow than a smaller winter rain burst which falls on saturated or frozen ground with less infiltration and higher percentage runoff.

10.18 The probability of occurrence of the *n*-year rain burst. A frequency analysis gives the probability of occurrence, or the return period, of rain bursts of designated magnitudes. It is of value to know the chances that a rain burst of at least this magnitude, or of a return period of n years, will occur within the next n years. For example, there is a probability of $1/n$ that the n year event will occur in a particular year and $1 - 1/n$ of it not occurring. Therefore the probability of it not occurring in n years is $(1 - 1/n)^n$, or its probability of occurring is $1 - (1 - 1/n)^n$. Thus the probability that the once-in-ten-yr rain burst will occur at least once in the next 10 yr is $1 - (1 - 1/10)^{10} = 1 - 0{\cdot}35 = 0{\cdot}65$. That is, there are 65 chances out of 100 that the once-in-10-yr rain burst will occur in the next 10 yr. A useful approximation to $1 - (1 - 1/n)^n$ is $1 - e^{-1} = 0{\cdot}63$ when n exceeds ten.

By further developing this probability theory the relationship between the design period, T of the rain burst and the probability that this rain burst will not be exceeded in a certain period T_p, which can be called the probability return period, is $P = 1 - (1 - 1/T)^{T_p}$. This is evaluated and shown in Fig. 10.16. [22] From this diagram, it is possible to calculate the risk, or probability, that a once-in-T-years rain burst will not be exceeded in a given number of years, T_p . Alternatively, if it is necessary to be 0·80 certain that a rain burst will not be exceeded in the next 10 yr, the return period of such a burst from Fig. 10.16 is obtained by proceeding from the design period 10 yr on the horizontal axis, vertically to the 80% probability line and then across to the probability return period of approximately 50 yr. This means that a once-in-50-yr rain burst has a probability of 80% of not being exceeded in the next 10 yr. If this is the desired risk then the value of rain burst selected is the once-in-50-yr rain burst.

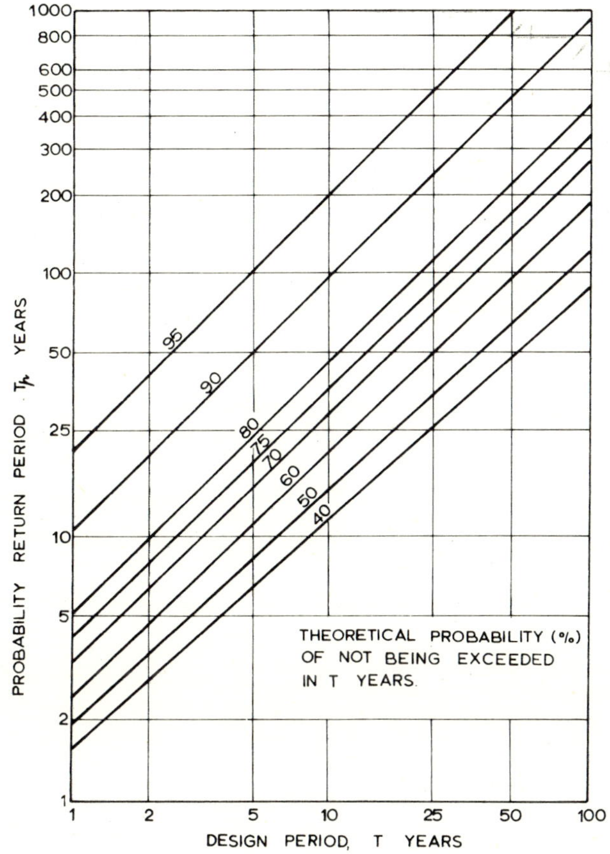

Fig. 10.16. Relationship between design period T and probability of this not being exceeded in period T.

REFERENCES

[1] WARNICK C.C., (1956). The wind field over rain and snow gauges. *Bulletin No. 11*, University of Idaho.

[1a] WARNICK C.C., (1953). Experiments with wind shields for precipitation gauges. *Trans. Am. Geophys. Union.* **34**, 379–388.

[2] STOREY H.C. and HAMILTON E.L., (1943). A comparative study of rain gauges. *Trans. Am. Geophys. Union.* **24**, 133.

[3] Bureau of Meteorology, Australia (1963). Report on development of a flood forecasting system for the lower Macleay River Valley, N.S.W., *Com. Bur. of Met. Melb.*

[4] U.S. Weather Bureau (1947). Thunderstorm rainfall. *Hydrometeorological report No. 5.* U.S.W.B.

[5] World Meteorological Organisation (1965). Guide to hydrometeorological practices. *W.M.O. Geneva.*

[6] BATTAN L.J., (1964). *Radar Meteorology*, The University of Chicago Press, Chicago.

[7] GUMBEL E.J., (1958). *Statistics of Extremes*, Columbia Univ. Press, N.Y.

[8] BROOKS C.E.P. and CARRUTHERS N., (1953). *Handbook of Statistical Methods in Meteorology*, H.M.S.O., London.

[9] FOSTER E.E., (1949). *Rainfall and Runoff*, Macmillan.

[10] CONRAD V., (1944). *Methods in Climatology*, Cambridge, Mass.

[11] CORN H.N., (1955). Rainfall depth, duration relationships. *Proc. Am. Soc. Civ. Engrs*, 81 paper, 840.

[12] SHANDS A.L. and BRANCATO G.N., (1947). Applied meteorology, mass curves of rainfall. U.S.W.B. *Hydromet. Tech. Paper No. 4.*

[13] BODY D.N., (1959). Flood estimation. *Water Research Foundation of Aust. Bull. No. 4*, Sydney.

[14] HERSHFIELD D.M., (1962). Extreme rainfall relationships. *J. Hydraulics Divn of Proc. Am. Soc. Civ. Engrs. Hy.* 6 73–92.

[15] O'MAHONY G., (1964). Methods of analysis of hydrological data. *Water Resources: Use and Management*, ed. WIESNER C.J., Melbourne University Press, Melbourne.

[16] Stormwater Standards Committee (1958). Australian rainfall and runoff. *The Institution of Engineers, Australia.*

[17] SHERMAN C.W., (1931). Frequency and intensity of excessive rainfalls at Boston, Mass. *Trans Am. S.C.E.* **95**, 951.

[18] HERSHFIELD D.M., (1961). Rainfall frequency atlas of the United States. *U.S.W.B. Tech. Paper, No. 40.*

[19] WEISS L.L., (1957). A nomogram for log–Normal frequency analysis. *Trans. Am. Geophys. Union* **38**, 33–37.

[20] REICH B.M., (1963). Short duration rainfall intensity estimates and other design aids for regions of space data. *J. Hydrology, Amsterdam* **1**, 3–28.

[21] U.S. Army Engineer School (1961). Student reference drainage. *E.O.11.*

[22] COURT A., (1952). Some new statistical techniques in geophysics. *Advan. Geophys, Vol. 1*, 45–85, Academic Press, N.Y.

CHAPTER ELEVEN

Precipitation and its relationship with space

11.1 Variation of precipitation with area. Registrations from rain-gauges represent extremely small samples of the rainfall over an area. For example, the catchment of an 8 in. dia. rain-gauge is approximately one eighty millionth of a square mile and, as gauges are usually separated by some miles, there is a high degree of extrapolation in using the gauge measurement as an indication of depth of a storm. However, the averaging out over an area of the meteorological properties and vertical lift in the storm system and its usually regular movement over a region, result in consistent amounts being deposited in well sited gauges during the storm period and give a homogeneous rainfall pattern.

In hydrological problems, the rainfall volume is required, or the product of rainfall depth over an area. It is convenient to use special methods to obtain the average depth during the period over the region concerned. The accuracy of rain-gauges in the assessment of rainfall depths has already been mentioned in paragraph 10.3 and depends upon the density and distribution of the rain-gauges, the size, nature, and topography of the catchment, the type of storm producing rainfall including its duration and movement, and the methods adopted to evaluate the data. The last is considered here, firstly using a single gauge and then using a network of gauges.

11.2 Extrapolating gauge rainfall for area. The single gauge is converted to an average depth of rainfall over an area by using empirical relationships established from dense networks. Figure 11.1 gives the one established by the United States Weather Bureau for use with rainfall frequency data. [1] The variation of the average rainfall depth with area is shown as a percentage of the rain-gauge fall. A different relationship is seen for each duration with steep gradients for the short durations. The diagram may be applied to rainfalls of all return periods and to continental areas similar to North America. It cannot take account of topographical features and holds for flat areas. Rain-gauge falls are converted to the average depth over areas up to 400 square miles by multiplying by the percentages read off Fig. 11.1 for the required duration and area. For example, a once-in-2-yr 6-hr fall of 2 in. can be extrapolated to an area of 300 square miles by multiplying by 0·84. An average depth of $2 \times 0{\cdot}84 = 1{\cdot}68$ in. is the estimate over 300 square miles. Extrapolation beyond 400 square miles is a doubtful process, although some relationships have been established and used beyond this range.

Formulae relating the average rainfall, $\bar{P}$, over an area A, for a given duration, to the point rainfall depth at the focus, P_0, have been established. This is equivalent to finding the equation of the average depth, area curve and is of the general form $\bar{P} = P_0 e^{-kA^n}$ where k and n are constants for a given storm or for an envelope of storms of given

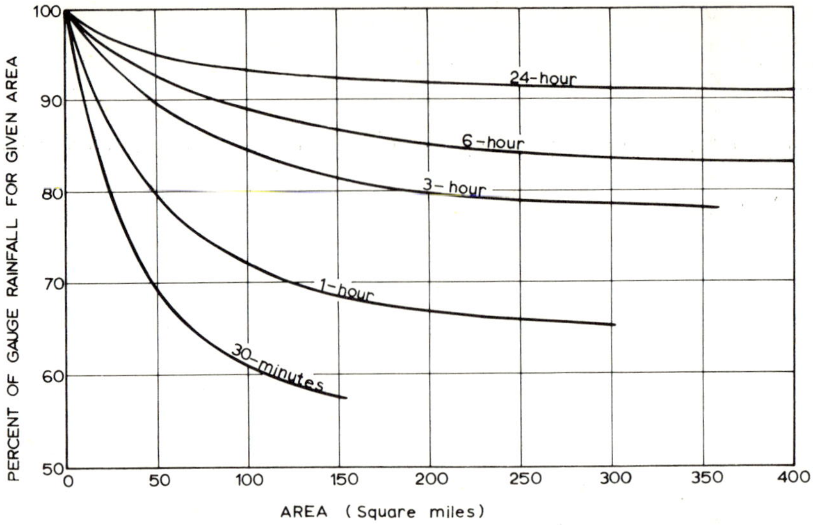

Fig. 11.1. Frequency rain-gauge readings related to area.

duration. By means of analysed storm data of focal rainfall depth for two areas and their equivalent average depths, the constants can be determined. For example, the envelope of five northern U.S. storms of 24-hr duration [2] is given by

$$\bar{P} = 16e^{-0{\cdot}0883A^{0{\cdot}24}}$$

and of five southern storms by

$$\bar{P} = 22e^{-0{\cdot}112A^{0{\cdot}23}}$$

Investigations over small areas in the United States [3] with dense networks suggest a relationship of the form

$$\bar{P} = a - bA^{1/2},$$

where a and b are constants which represent the maximum rainfall at the focus and the mean rainfall gradient. A similar relationship involving the square root of the area was used by Fletcher in a formula fitting the maximum depth, duration, area data of the United States. This is discussed in paragraph 18.6.

If storm isohyets were circular and the intensity fell off linearly it has been shown that the average intensity is one third of the focal intensity. [4] Very few natural storms conform to these specifications and the empirical relationships above better describe the storm's variations. Extreme types of storms such as cloud bursts have their own distinctive areal pattern which is shown in Fig. 18.2. Intense tropical cyclones or hurricanes also show some differences in areal pattern from the more common types. However, for the extrapolation of frequency derived rain-gauge data to larger areas, Fig. 11.1 is the most satisfactory.

Topography and local meteorological factors such as the distance from the moisture source, or storm type, distort the storm's areal pattern and extrapolation is therefore restricted to relatively flat areas within meteorologically homogeneous zones. Extrapolation is uncertain beyond 400 square miles and, for larger areas, a network of rain-gauges, or information on topography and meteorology of the region, is needed to synthesis the areal pattern.

11.3 Rain-gauge networks converted to average rainfalls over an area. The average rainfall over an area is most conveniently determined from a well-sited network of rain-gauges which show the local variations of rainfall. The importance of setting up representative networks of rain-gauges as soon as possible cannot be over-emphasised. The principal methods of converting the individual gauge readings to an average depth over the catchment are as follows:

Arithmetic mean.
Areal weighting or 'Thiessen'.
Triangulation.
Isohyetal with linear interpolation.
Isohyetal with subjective interpolation.
Weights based on isohyetal pattern or other data.

11.4 Arithmetic mean method. The arithmetic mean of the stations in the network is the simplest method and involves taking the arithmetic mean of the rainfall stations within the catchment.

$$\bar{P} = \frac{P_1 + P_2 + \ldots + P_n}{n} = \frac{\sum_1^n P_n}{n}. \tag{11.1}$$

Where topography is reasonably flat, gauges are evenly spaced, and there are only small variations in rainfall, this gives accurate estimates of rainfall depth. In other areas, a network selected by trial and error or, by personal judgement and local knowledge, to fully represent the catchment as described in paragraph 10.3, can give satisfactory results. Each gauge is given equal weight, and in a select network, the deficits of one gauge may balance the surpluses of others. This method gives a quick and objective result and is most suited to machine or computer operation.

11.5 The areal weighting method. The 'Thiessen' or areal weighting method, weights each station according to the area which is closer to that station than to any other. This area is found by drawing the perpendicular bisectors of the lines joining nearby stations so that a polygon is formed around the station. The area of each polygon is found, by planimetering or other means, and becomes the 'Thiessen' weight for the station in the centre of the polygon. One polygon is drawn for each station. Each catchment, with its rain-gauge network, has its set of polygons and corresponding 'Thiessen' weights. If the network changes because one station does not report or a new rain-gauge is added, new polygons and weights have to be determined for the new network. If a few observations only are missing from a network, it is more convenient to estimate the missing data, by methods to be later discussed, than to determine the new weights. In constructing the polygons, near equiangular figures should be sought, the closest stations used first, and re-entrant angles avoided. Stations outside the catchment may be used in this method with an appropriate weight using only that portion of the polygon inside the catchment. The method assumes that precipitation varies linearly between stations, and allows for non uniformity of the gauge network but not for topographical or meteorological factors.

If $P_1, P_2, P_3, \ldots, P_n$ represent the precipitation at stations 1, 2, 3, ..., n, and the areal weights are $A_1, A_2, A_3, \ldots, A_n$, then the mean depth of precipitation $\bar{P}$ is

$$P = \frac{A_1P_1 + A_2P_2 + A_3P_3 + \ldots + A_nP_n}{A_1 + A_2 + A_3 + \ldots + A_n} = \bar{P} = \frac{A_1}{A}P_1 + \frac{A_2}{A}P_2 + \frac{A_3}{A}P_3 + \ldots + \frac{A_n}{A}P_n$$

If weighting factors

$$W_1 = \frac{A_1}{A}, W_2 = \frac{A_2}{A}, W_3 = \frac{A_3}{A}, \ldots, W_n = \frac{A_n}{A}$$

Then $$\bar{P} = \sum_1^n W_n P_n, \tag{11.2}$$

or if W is in percentages $$\bar{P} = \frac{\sum_1^n W_n P_n}{100}. \tag{11.3}$$

An example of the use of this method is shown in Fig. 11.2.

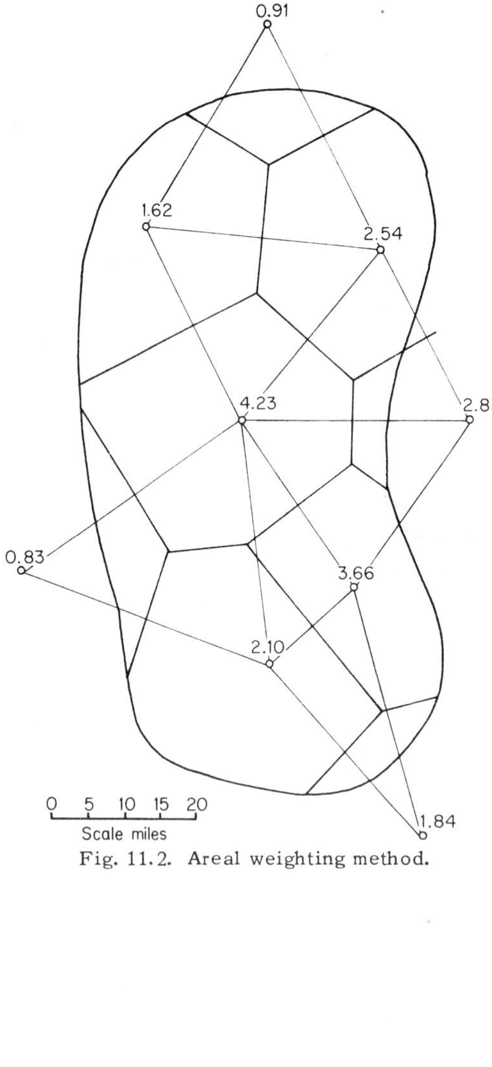

Fig. 11.2. Areal weighting method.

ARITHMETIC MEAN

$$\frac{1{\cdot}62 + 2{\cdot}54 + 4{\cdot}23 + 3{\cdot}66 + 2{\cdot}10}{5} = 2{\cdot}83 \text{ in.}$$

AREAL WEIGHTING

PRECIP. in.	AREA units	AREA % total	WEIGHTED P in (1)×(3)÷100
1·62	6·10	16·27	0·2635
4·23	8·36	22·27	0·9440
0·83	1·65	4·40	0·0365
2·10	7·53	20·10	0·4220
1·84	0·84	2·24	0·0412
3·66	4·70	12·53	0·4585
2·83	0·73	1·95	0·0551
2·54	6·16	16·42	0·4170
0·91	1·41	3·78	0·0342
Total	37·48	99·96	2·67

Although it is convenient to use percentages or areas in square miles, the actual planimeter readings may be used as weights provided the sum of these are also used to find the average depth over the catchment.

11.6 The triangulation method. This method is similar to the areal weighting method and consists of constructing a series of triangles by joining pairs of adjacent rainfall stations to form a series of triangles. The nearest stations should always be joined and the triangles kept near the equiangular shape when possible. The mean depth of rainfall over each triangle is the mean of the three rainfall registrations at its apices. This is equivalent to using station weighting factors equal to one third of the areas of the triangles surrounding the stations. These weights may therefore be determined and applied in the same way as in the areal weighting method of 11.5. Figure 11.3 is an illustration.

11.7 The isohyetal method–linear and subjective interpolation. The rain-gauge network may be used to construct isohyets or contours of equal rainfall depth. All places along the isohyets experience the same rainfall. The positioning or drawing of the isohyets may be decided in several ways. Firstly, pairs of stations may be used to linearly interpolate the rainfall to give the location of the whole number rainfall depth. Equal points determined for the network may then be joined by straight lines to give a formal-looking pattern as in Fig. 11.3. The isohyets are thus located objectively and there should be little variation between different workers using the same data. There may be alternative isohyets drawn by selecting different pairs of stations, but if the station with the steepest gradient is used first, the method becomes consistent. The depth of rainfall over the catchment is found by multiplying the area between the isohyets by their mean value. It can be shown that the linearly interpolated map gives the same result as triangulation. The data normally used does not warrant this precise interpolation, which is both time consuming and tedious.

The second and more common isohyetal method is to use personal knowledge and skill in interpolating the data. Allowances can be made for topography, distance from sea, mean annual, seasonal, or monthly rainfall pattern, and for storm incidence and frequency. Any known variation of rainfall with height, preferred paths of storms, spillover, and rain shadow effects can all be allowed for in the subjective drawing of the isohyets. If desired, linear interpolation may be made by the analyst by approximate methods or by eye. Although it is possible for different results to be achieved by different workers, there is seldom any large difference, provided all available data is used.

Using this method the precipitation over an area is computed by planimetering the areas between the isohyets and multiplying them by the mean values of the precipitation between these isohyets. With straight isohyets and near linear variation of precipitation, the arithmetic mean of the isohyets is an accurate assessment of the mean value, but with marked curvature and irregularity in rainfall gradients, it is better to assess the mean depth in accord with the available facts. Refinements of the method have been devised, but are generally unnecessary and do not justify the greatly increased work.

The subjective isohyetal method has the advantage that all data and knowledge of the area can be used and the result clearly displayed. Any anomalies in the rainfall pattern can be explained, corrected, queried, or discounted. Its disadvantage is that considerable

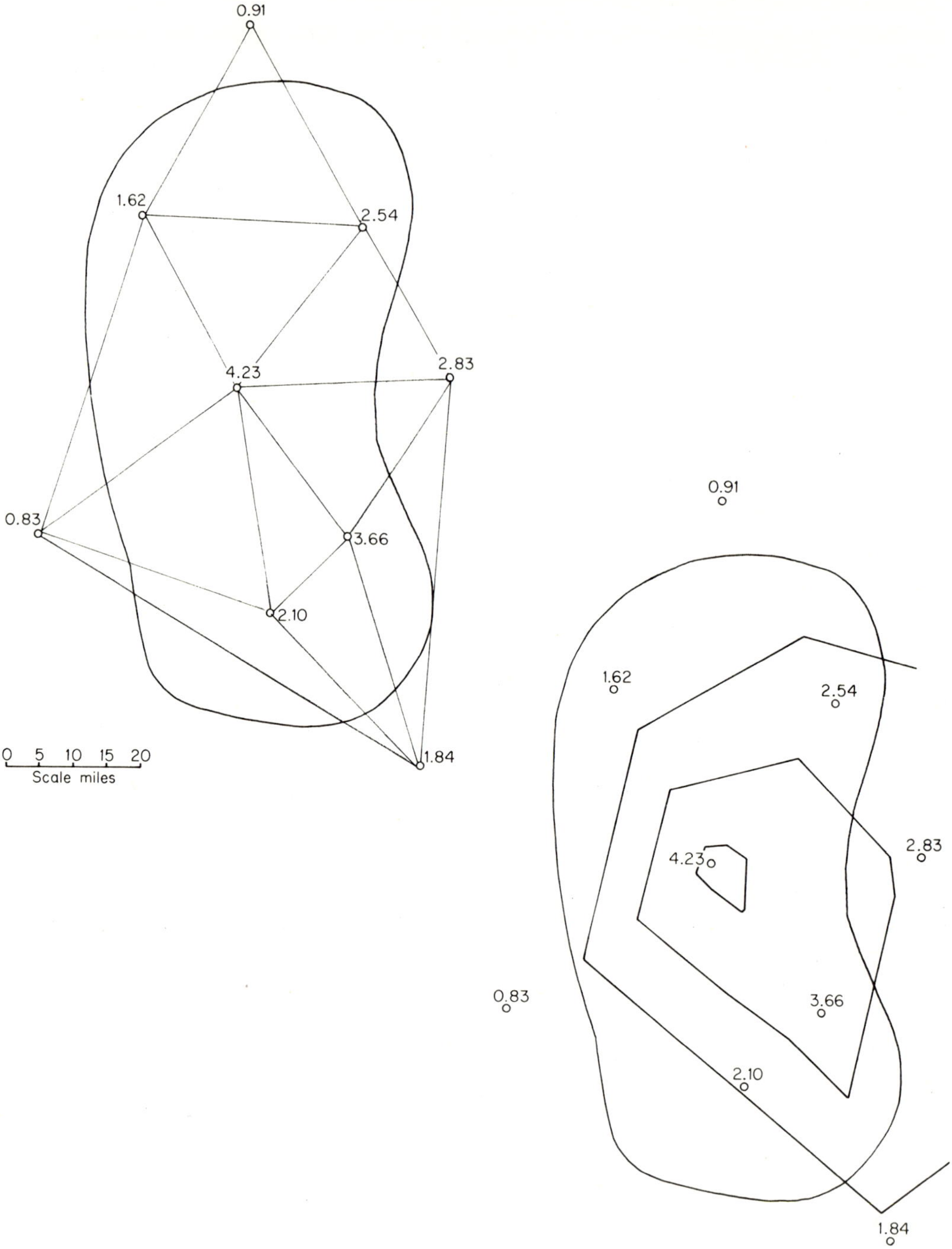

Fig. 11.3. Triangulation and linearly interpolated isohyetal method.

time is employed in constructing maps for each situation and assessing the areas and depths. There is also the possibility that the analyst may, unconsciously, be biased and unduly weight one particular area or zone when constructing his isohyetal map. Figure 11.4 shows the same data as Fig. 11.2 assessed by the subjective isohyetal method.

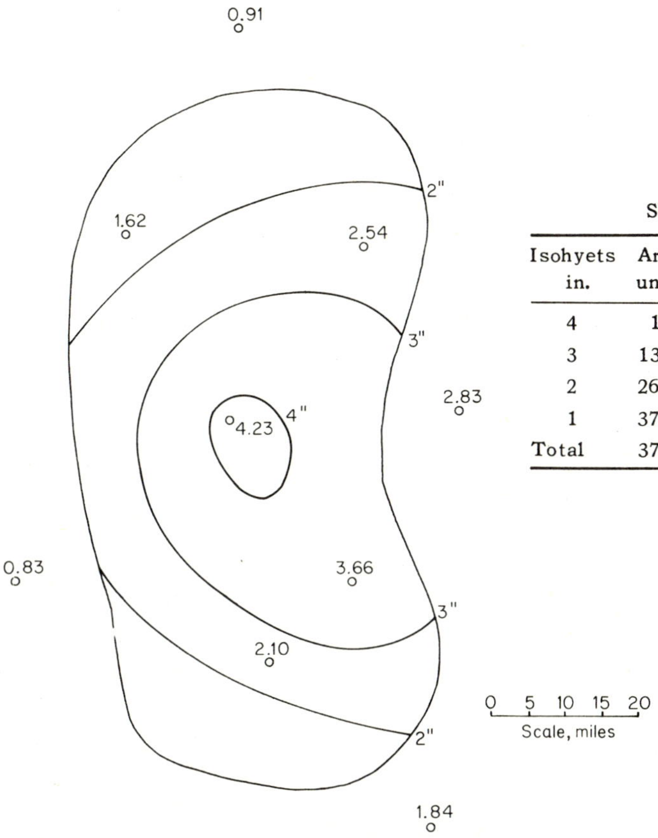

Fig. 11.4. Subjective isohyetal method.

SUBJECTIVE ISOHYETS

Isohyets in.	Area units	Area net	Area %	Av. precip. in.	P (4)×(5)÷100
4	1·11	1·11	2·96	4·1	0·121
3	13·48	12·47	33·00	3·5	1·155
2	26·58	13·10	34·93	2·5	0·873
1	37·50	10·92	29·11	1·5	0·448
Total	37·50	37·50	100·00	11·6	2·607

11.8 Weights based on isohyetal patterns or other data. The isohyetal method is generally recognised as giving the most accurate representation of the areal pattern of precipitation. However, if numerous computations of rainfall depths are to be made it is more time consuming than the areal weighting method once weights have been decided. Also, the latter is applicable to machine methods. It is possible, instead of weighting stations for area only, to take account of other factors such as the average rainfall pattern, the amount of runoff, or storm incidence. These weights can be established by theoretical, empirical or historical considerations, or a combination of them. Weights may be adjusted to allow for the rainfall incidence in the catchment. For example, if the average annual isohyetal map is drawn over the normal 'Thiessen' network it is possible to evaluate the depth of average annual rainfall within the station polygon. If this agrees with the annual average at the station, the areal weight is unchanged. However, if it is higher, or lower, say 110% or 90% of the station depth, the areal weight is increased or decreased by the same percentage. Higher percentages at some stations may be balanced by lower ones at others. Using this method, it follows that if the average isohyetal map over the catchment gives a higher or lower depth than the 'Thiessen' evaluation, the overall weights are to be increased or decreased by the same percentages and the revised weights will more nearly represent the catchment conditions. Also, weights can be based on the stream flow characteristic of an area. If each rainfall station has a corresponding stream gauging site, the percentage of the total stream flow out of the catchment which each station contributes may be used as a weighing factor. If there are a number of rainfall stations to the stream gauging site, the arithmetic mean of these stations might be multiplied by the weight.

Another method of arriving at weights involves using isohyets of a common storm pattern or the average annual rainfall, and starting from the centre of heaviest rainfall, plot the precipitation depth against the percentage of catchment area it represents. This gives a curve of percentage catchment area against precipitation depth as shown in Fig. 11.5.

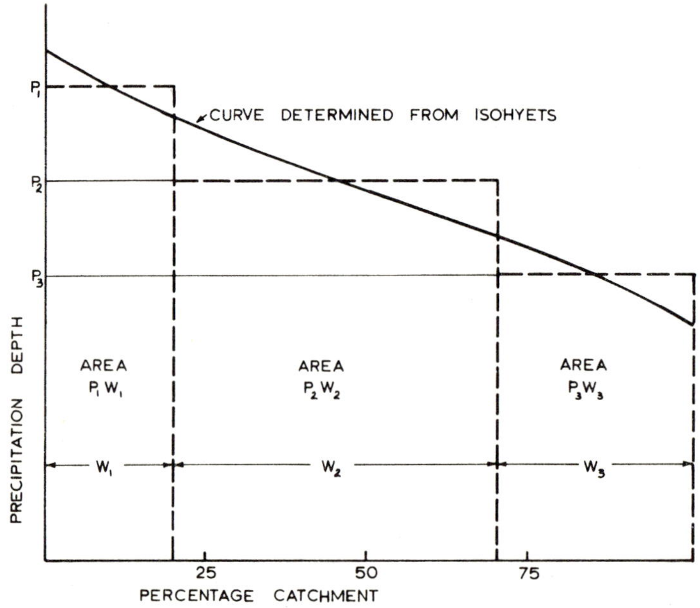

Fig. 11.5. Weights determined from area-depth curve.

If horizontal lines, representing the rainfall at each station, are now drawn to cut this curve, blocks may be drawn such that their area is approximately equal to the area under the curve. Each block will now represent the contribution of that gauge to the total volume of rainfall over the basin. The width of this block in percentage is the weight for that gauge. Using Fig. 11.5, if three rainfall stations report P_1, P_2 and P_3 over the period, their weights will be W_1, W_2 and W_3. This method may be used to simplify rainfall depth calculations by determining the percentage area depth curve from a dense network,

Table 11.1. *Comparison of methods which assess rainfall from a network.*

	Arithmetic mean	Areal weighting polygon	Triangulation	Isohyetal		Revised Weighting
				Linear	Subjective	
Nature of method	Objective	Objective	Objective	Objective	Subjective	Subjective assessment Objective in use
Speed	Rapid	Fast when weights known	Fast when weights known	Very slow	Slow	Fast when weights known
Effect of missing data	No allowance reduced accuracy	No allowance Reweight or extrapolate	No allowance Reweight or extrapolate	Estimate missing data	Estimate data easily	No allowance, reweight or extrapolate
Allowance for physiography	None	None	None	None	Yes	Some allowance
Display of data and correction of errors	Poor	Poor	Poor	Good	Excellent	Poor
Edge or boundary errors	None	Allowance made	Some allowance difficult with curves	Unsuitable	Satisfactory	Satisfactory
Adaption to machine processes	Excellent	Good	Good	Poor	Poor	Good
Accuracy	Satisfactory with even rainfall good network and flat country			Excellent in most circumstances		Satisfactory in most conditions

and then selecting a few representative stations and weighting them suitably.

11.9 Comparison of methods. The accuracy of the methods described from 11.4 to 11.8 depends upon the nature of the rainfall, the rain-gauge network, and the physiography of the catchment. Storms with evenly distributed rainfall, over flat country where there is a well spaced network, give similar results for all methods and the simpler ones are to be preferred. Complex weather situations, poorly distributed rain-gauges or catchments of varied topography are more accurately assessed by the isohyetal method or the revised weighting method as there is scope for skilled assessments. Table 11.1 gives a comparison of the methods and the factors involved.

11.10 Depth, duration, area analysis of storms. The variations of storm precipitation with duration and area have been discussed separately but in most applications, the depth over a known area for a given duration is required. Where possible, the frequency of the event, or its return period, should also be known. The methods so far discussed show that rain-gauge registrations can be analysed to give depths over a given duration and for a prescribed return period and then, extrapolated to areas up to 400 square miles by empirical relationships. Networks of rain gauges however, give a better representation of the storm and over larger areas. By means of isohyetal maps and mass curves, the maximum depths of precipitation yielded by the storms may be determined for given durations and areas. In other words, each storm may be expressed in terms of the maximum depths which it gives over a stated area and for a given duration. This is termed the depth, duration, area data of the storm. It might be noted that a storm may yield for a period of 6 hr, say, from 6 a.m. to noon, an average depth of 5·1 in. over 1000 square miles, but an examination of the whole duration of the storm may show that some other period, say, 11.15 a.m. to 5.15 p.m. gives the highest amount of 6·7 in. for 6 hr over an area of 1000 square miles. The maximum 6-hr, 1000 square mile depth for that storm is thus 6·7 in. A comparison with all other storms in Table 18.3 gives the greatest depths which have occurred in the United States. For 6 hr and 1000 square miles this is 13·4 in.

Depth, duration, area data represent a useful form of expressing a storm's potentiality for producing runoff. They may be extrapolated to their probable maximum values for use in flood estimation studies. Return periods of depth, duration, area values are not generally available and are difficult to estimate. The ratios of the probable maximum 6 hr 10 square miles depths, to the once-in-100-yr 6-hr depth have been estimated in the U.S. and show geographical variations from less than 2 to 9. [1a] Apart from this and the work on statistical methods in paragraph 18.8, there is little information on return periods of depth, duration, area data.

The depth, duration, area data of a storm can be expressed in the form of a table, of which Table 11.2 is an example.

Graphically, this data may be shown in Fig. 11.6 as a family of depth-area curves where the linear depth is plotted against logarithm of the area and the members of the family are identified by duration. Alternatively, a family of depth duration curves shown in Fig. 11.7 may be constructed with the linear depth plotted against the logarithm of the duration with the members of the families specified by area.

Table 11.2. *Maximum depth, duration, area precipitation (in) of a storm*

Area	Duration (hr)			
(square miles)	12	24	48	72
100	8	13	16	18
500	7	12	15	17
1000	6½	11	14	16
10000	4½	8	11	13

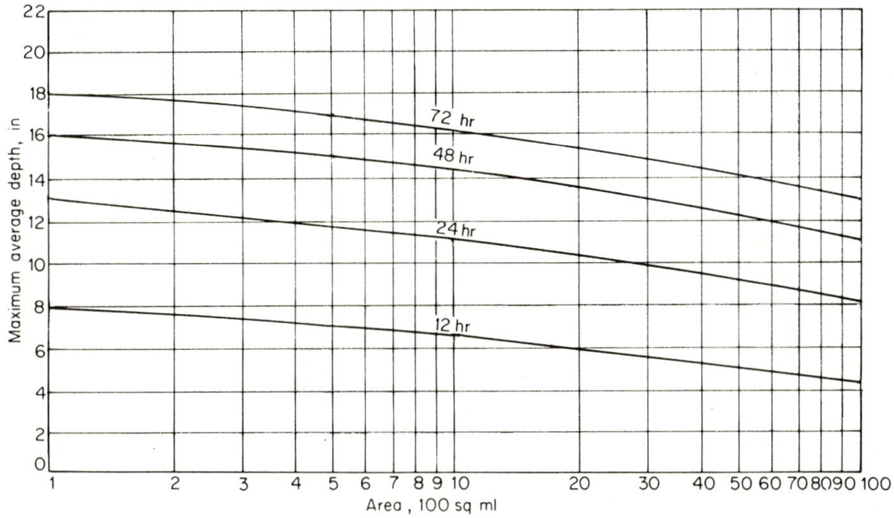

Fig. 11.6. Family of depth-area curves.

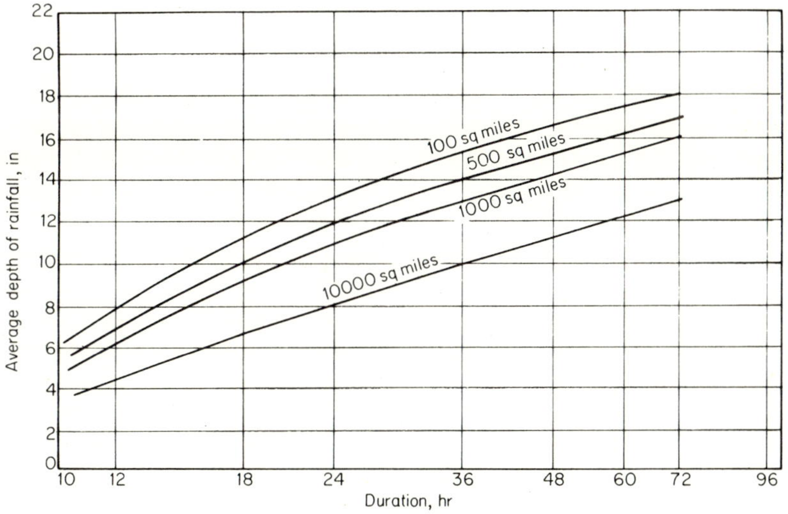

Fig. 11.7. Family of depth-duration curves.

Techniques have been developed to obtain the maximum depth, duration, area data of storms and demand a large amount of careful checking, compiling, and assembling of data from all available sources, published and unofficial. These techniques are fully described in publications. [5–7] The principle steps are as follows:

(*a*) A preliminary analysis indicates the storm period and area of interest. It leads to a complete collection, interpolation, and check of all data by the various techniques described here and in the publications.

(*b*) Mass curves of rainfall are constructed from pluviograph records or synthesised from synoptic reports, unofficial records, and a knowledge of storm mechanisms. The use of transparent graph paper ruled at hourly intervals permits easy comparisons and interpolations. The time period of the storm and the intervals of dissection are decided. For the large storms, 6-hr intervals are used, but for smaller storms of short duration, less than 500 square miles and 24 hr, an interval of 1 hr or 15 min may be needed. The purpose for which the data is to be used also controls the interval used. For example, an urban drainage problem requires short duration falls.

(*c*) A detailed isohyetal map is constructed for the storm period. The various storm or principal rainfall centres are then apparent and are the bases for establishing zones. In each zone, the isohyets are planimetered to give rainfall depths over corresponding areas. The zones are combined for larger areas by assuring that all the areas combined are encompassed by the same isohyet.

(*d*) The time pattern of the rainfall is given by the mass curves of each zone and is represented by the increments of rainfall in each interval. Each of the mass curves may be weighted by their 'Thiessen' polygons to better show their sphere of influence. The average rainfall depth over the area encompassed by a given isohyet can be allotted a time pattern by using the 'Thiessen' weights for the mass curves within that isohyet and preparing a weighted mass curve. This can be adjusted to the depth within the isohyet. From this tabulation the maximum rainfall depths for the unit interval and multiples of it can be obtained and listed against the corresponding area. This process is continued for other encompassing isohyets within the zone or significant combinations of them. The highest depths for each duration and area are now apparent and all data is plotted on a depth–area diagram with the third factor, duration, written against the plotted point. An enveloping curve, can now be drawn to the depths of a given duration and repeated for the other durations. Figures 11.8, 11.9, and 11.10 show the isohyetal map, zones and 'Thiessen' polygons, and the maximum depth, duration, area values for an Australian storm.

A simple numerical example can assume that the 5-in. isohyet encompasses an area of 550 square miles with an average depth of 7·2 in. and surrounds three pluviographs A, B and C which have 'Thiessen' weights of 50, 40 and 10 within the 5-in. isohyet. Table 11.3 shows the method of weighting the mass curves and deducing the maximum depth, duration values for an area of 550 square miles.

If there are a large number of pluviographs, 6 or more, within an isohyet, the arithmetic mean of the various mass curves is satisfactory or each pluviograph exerts the same weight.

As an alternative method, a series of isohyetal maps for say 6, 12 etc. hr may be drawn for the more intense periods of the storms. The intervals of maximum intensity are determined from the pluviographs or mass curves. These isohyetal maps may be planimetered to give the maximum depths over the various areas for the intervals of the maps. This method is easier to use especially for simple storms in which the period of intense rain is clearly defined.

Table 11.3. *Weighting mass curves within 5-in. isohyet.*

Time	Total fall 3 a.m.–9 a.m. Depth	Weighted	Total fall 3 a.m.–3 p.m. Depth	Weighted	Total fall 3 a.m.–9 p.m. Depth	Weighted	Total fall 3 p.m.–3 a.m. Depth	Weighted
Pluvio A	3·0		4·0		5·0		8·0	
× 0·5		1·5		2·0		2·5		4·0
Pluvio B	2·0		3·0		4·0		6·0	
× 0·4		0·8		1·2		1·6		2·4
Pluvio C	2·0		2·0		3·0		5·0	
× 0·1		0·2		0·2		0·3		0·5
Average mass curve within 5 in. isohyet		2·5		3·4		4·4		6·9
Adjusted to depth of 7·2 in. over 550 square miles $\times \frac{7\cdot 2}{6\cdot 9}$		2·6		3·5		4·6		7·2

Maximum depth for 6 hr 550 square miles = 2·6 in. from 3 a.m.–9 a.m. or 9 p.m.–3 p.m.
Maximum depth for 12 hr 550 square miles = 3·7 in. from 3 p.m.–3 a.m.
Maximum depth for 18 hr 550 square miles = 4·6 in. from 3 a.m.–9 p.m. or 9 a.m.–3 a.m.

11.11 Estimating missing precipitation records. [8,9] If precipitation records at a station are missing for a portion of the period it is better to estimate them rather than to leave a gap. This applies especially when data is processed with automatic equipment.

Missing precipitation records may be interpolated from the isohyetal maps which cover the period in question. Skilfully drawn maps take local rainfall characteristics into account including the average pattern, distance from the sea, topography and storm paths. The time necessary to construct these maps is a disadvantage, but if the isohyetal method is to be used to evaluate the areal rainfall, additional time is not involved.

In cases where there is little variation, say, less than 10% in the average annual rainfalls between adjacent stations, the simultaneous rainfall registrations at the nearest surrounding stations may be averaged and used as an estimate of the rainfall at the missing station. Usually, three stations will give a good estimate.

When the average annual rainfalls in an area differ by more than 10% the method of ratios is preferred in estimating missing records. In this technique, the check or index stations are assigned weights, equal to the ratios of the average annual precipitation of the missing station to the average annual precipitation of the check stations. To obtain the precipitation at the missing station the precipitation depths during the period at the check stations are multiplied by the weighting factors and averaged. For example, for three index stations A, B, and C the estimated precipitation at the missing station X is

$$P_x = \frac{1}{3}\left(\frac{\bar{P}_x}{\bar{P}_A} P_A + \frac{\bar{P}_x}{\bar{P}_B} P_B + \frac{\bar{P}_x}{\bar{P}_C} P_C\right). \tag{11.4}$$

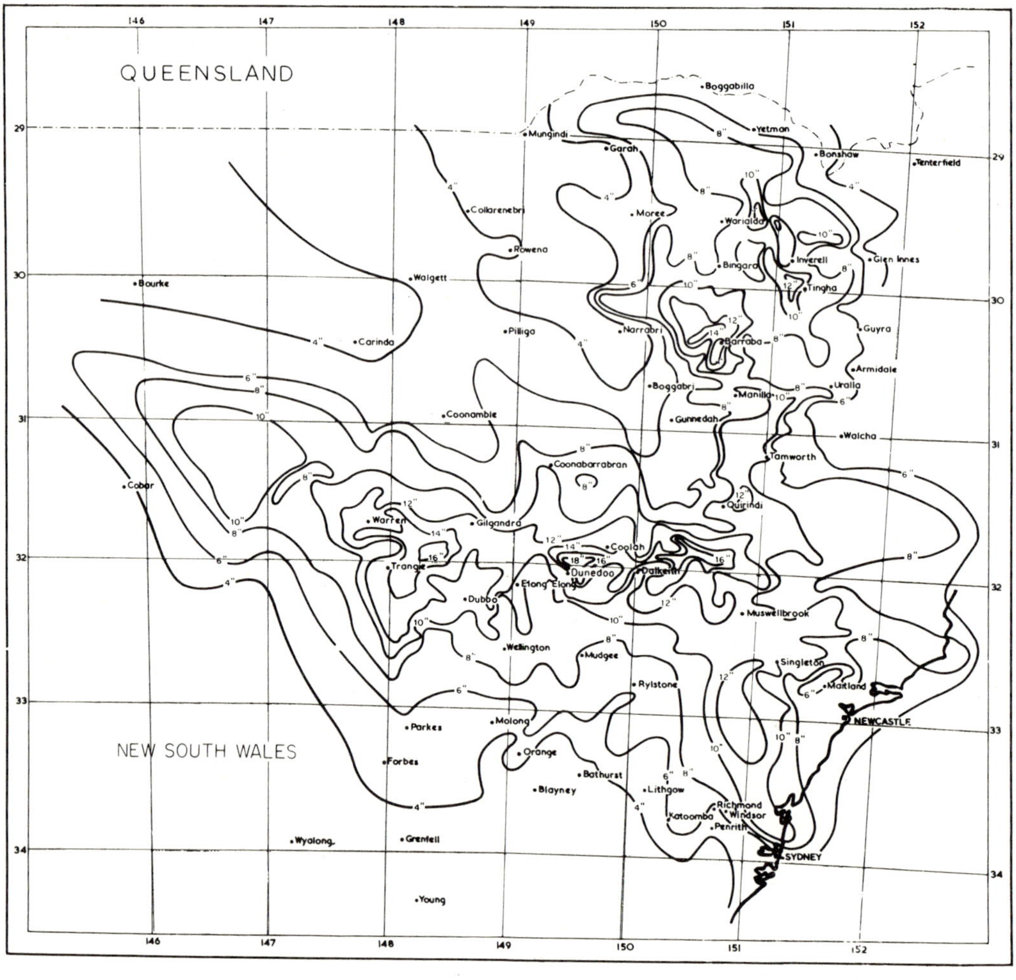

Fig. 11.8. Isohyetal map for an Australian storm.

There is little advantage in using many stations, in fact two or three index stations are adequate in most cases.

The method of ratios can also be used to obtain estimates of long term means at stations with short records. The short term average at station G, is compared by ratio with the simultaneous average at an adjacent station H, which also possesses a long term average. The estimated long term average for G is equal to

$$\frac{\text{the short term average for } G}{\text{the short term average for } H}$$

multiplied by the long term average for H.

The method of ratios is basically similar to the double mass curve technique where rainfall records are checked for consistency.

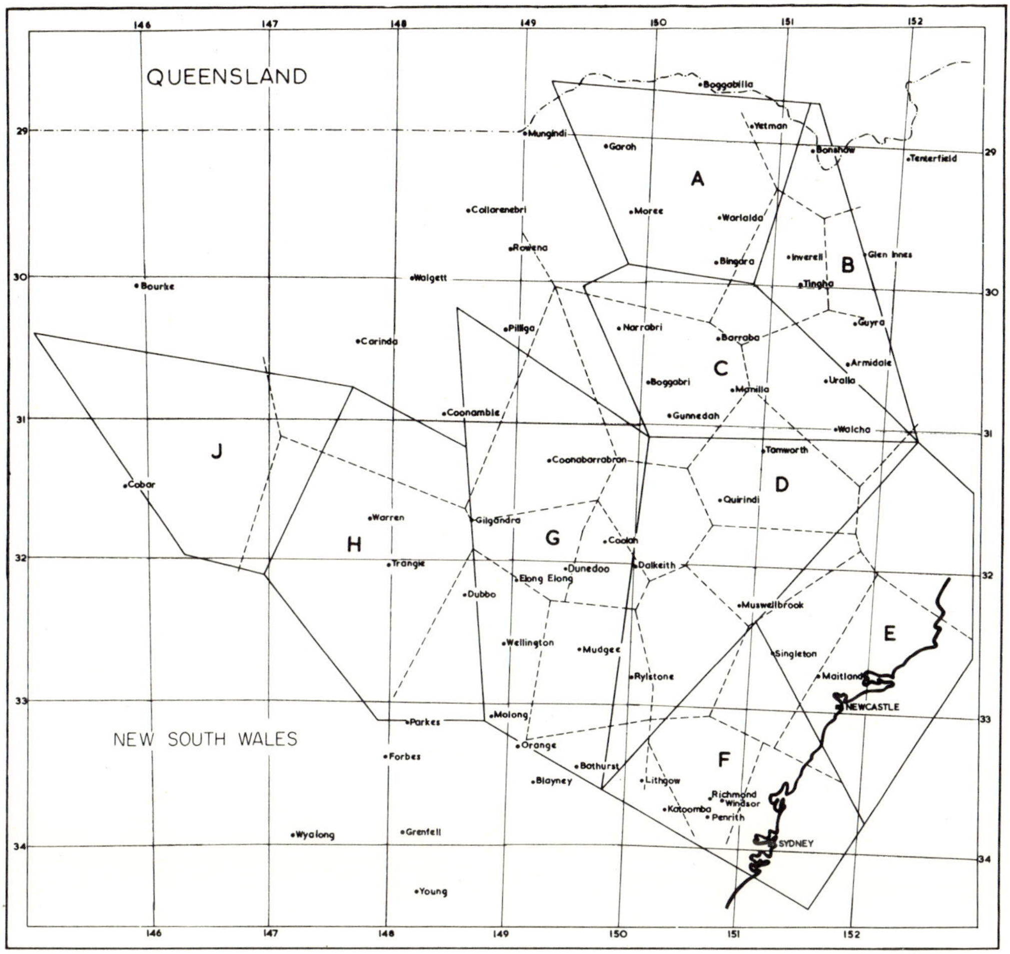

Fig. 11.9. Zones and 'Thiessen' polygons.

11.12 Analysing precipitation data by double mass curves. [10,11] Emphasis has been placed on the correct siting of rain gauges and the precautions necessary to obtain representative precipitation readings. Present rain gauge sites can be checked for these features, but in the past, different sites may have been used, or the exposure changed by the growth of trees and the construction of buildings. Also, there may have been a change in instruments, say from a 5-in. to an 8-in. rain-gauge. Again the rain-gauge may have been faulty for part of the time or the method and accuracy of measurement may have been suspect. Although a note is usually made in the station register of the known changes of

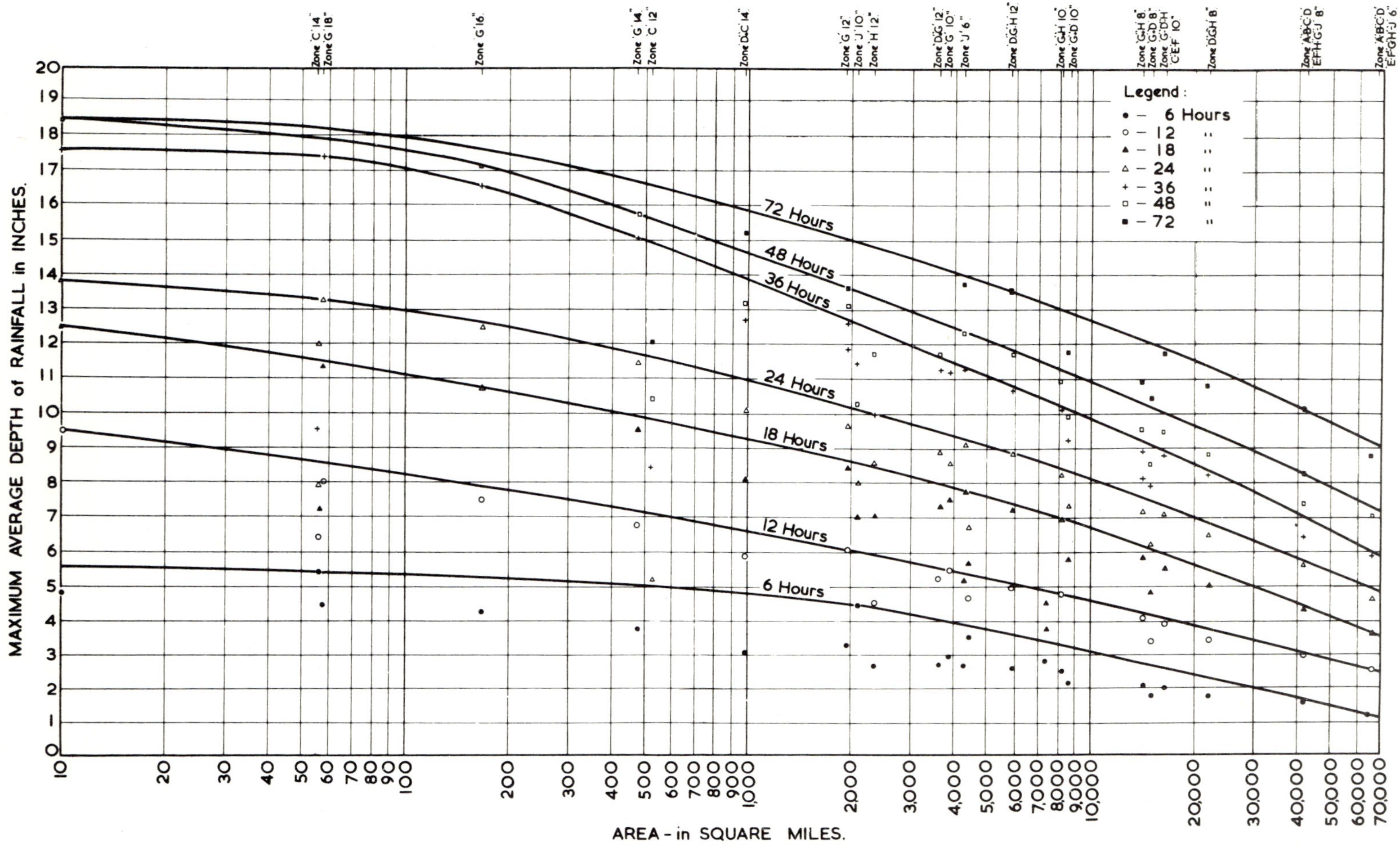

Fig. 11.10. Maximum depth, duration, area curves.

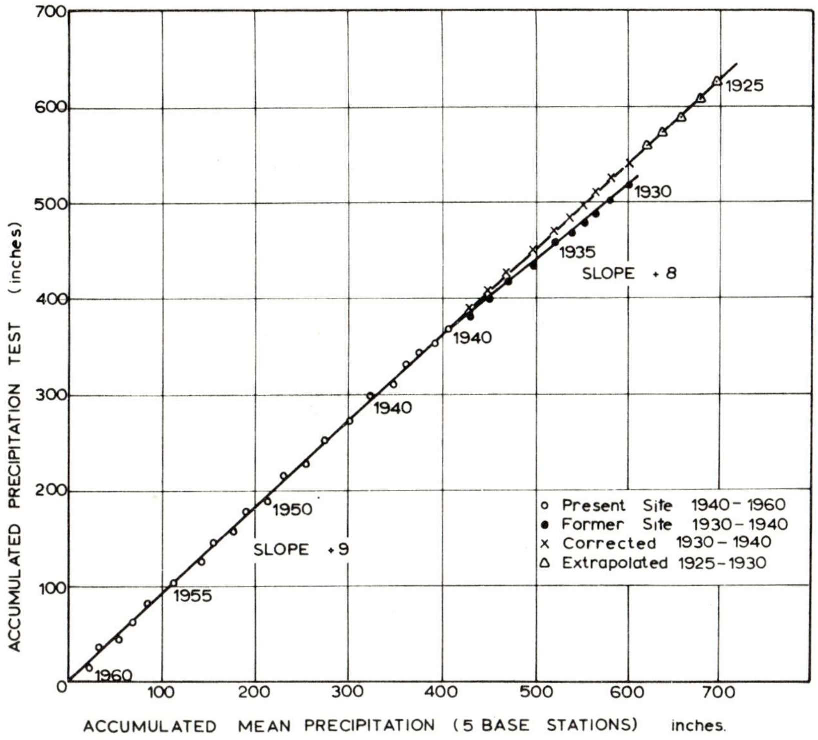

Fig. 11.11. Double mass curve.

site and instruments, their effect, or other unknown factors, may throw doubt on the accuracy and consistency of the record during the period covered. The double mass curve technique tests records for consistency and accuracy, and provides a correction factor to assure that the data is reasonably homogeneous throughout its period of record and related to a known site. It enables data to be estimated for missing periods or to be extrapolated beyond the existing length of record. It is based on the method of ratios discussed above.

The double mass curve is plotted by comparing the accumulated, daily, monthly, seasonal or annual rainfalls at the station under test with the corresponding accumulated depths of mean precipitation at a number of nearby stations. Experience has shown that an average of five stations give a stable mean, but fewer are sometimes used. It may be necessary to test stations against an assumed base for consistency so that the stations form a stable base. Inconsistent stations are either rejected or adjusted.

In constructing a double mass curve, it is convenient to start from the most recent data as details of the site and gauge are known. Also, a recently established station's records may need to be extrapolated to give estimates for earlier years.

Methods of correction and extrapolation are apparent from Fig. 11.11 where the double mass curve is marked by a change of site in 1940 shown by a change in the slope of the curve. For the period 1960–1940 the ratio of the accumulated precipitation at the test site to the accumulated mean precipitation at the base stations is 0·90. From 1940–1930 the ratio is 0·80. In order to correct the older records to be consistent with the more recent ones, those from 1940–1930 are multiplied by 0·90/0·80. If records from the base stations were available prior to 1930 but not at the test station, approximate estimates for these years may be obtained by continuing the line of slope 0·90 and, from the accumulated values at the base stations from years prior to 1930, read off the corresponding value expected at the test station. Subtracting adjacent values gives the depth for the intervening period. Alternatively accumulated mean depths prior to 1930 may be multiplied by 0·90. The presence of a growing tree near a gauge with an increasing sheltering effect would show on the double mass curve as a curved line of changing ratio.

The points comprising the double mass curve fluctuate about the line of best fit and any change of slope should be supported by at least five points or other evidence suggesting a change in precipitation measurement. Double mass curves are usually used for longer term data of at least a month in length. With sporadic storms or in places of rapid change of precipitation with distance as in mountainous areas, it should be used with caution.

11.13 Other methods of correcting precipitation data. Precipitation records may also be tested for consistency, corrected, and adjusted by the use of isopleths or by the examination of anomalies.

The plotting of data on maps and the drawing of isopleths is a good method of display, in which the geographical variations are obvious. A regular pattern should be apparent and the departures, especially erratic ones, can be logically explained. For example, mountain ranges show intensification on their windward side and a rain shadow effect on their lee side. Isolated storms may also affect few stations only. Any irregularity in the isopleths which cannot be explained should have the data in their vicinity checked for error and either corrected, rejected, or admitted with revision of the pattern. The isopleths from the accepted stations may be used to give a correction factor.

The average distribution of rainfall, or the average isohyetal map for the month, season or year, may be used as the most probable pattern for rainfall. Any unusual departures from this pattern may be detected and checked. The data over a given period may be expressed as a departure from the mean or as a fraction or percentage of the mean. The geographical distribution of these may show anomalies. For example, the rainfalls in a particular year may be expressed as percentages of the mean annual rainfalls. These percentages, plotted on a map, should show a consistent pattern which is best displayed by drawing isopleths or isopercentals. Isolated large deviations from the regular pattern should be checked. Percentages at nearby stations may give correction factors if needed and enable improved maps to be drawn.

Instead of selecting the average rainfall pattern as representative of normal conditions, rainfall of a certain frequency may be selected. The once-in-two-yr rainfall is often used as giving a likely basic pattern and departures from this are used in the same way as departures from the average annual rainfall.

If frequency analyses are available and are plotted on maps as isohyets of the rain bursts of various recurrence intervals, it is possible to use these to assess the approximate recurrence interval of the rainfalls being examined and assure that they are logically consistent. In making decisions, the nature and type of storm must be known and assessed. For example, isolated thunderstorms give far more irregular patterns than the regular distribution from synoptic storms which cover wide areas. Experience in the meteorology of the region is necessary.

11.14 Variation of precipitation with other factors. Besides its relationships with time and area, precipitation displays variations with other parameters. These may be physiographic, like height, slope, aspect, and distance from the sea, or dependent on air mass properties such as precipitable water, stability, and wind velocity. Precipitation intensity can be related to these parameters and empirical constants found from data in specific locations and storm situations. Some of these are discussed in later sections. The setting up of orographical storm models to compute the probable maximum precipitation in paragraph 16.16 and the establishment of empirical rainfall equations in the Willamette valley in paragraph 18.7 are typical examples.

On occasions, rainfall has been related to one particular variable but with limited success as in most situations, there is more than one significant variable. For example, an empirical relationship can be established between rainfall and elevation, an equation for New England being $P = 35 + 7{\cdot}5z$ where z is the altitude in thousands of feet. Similar relationships can be derived in localities where mountain ranges are athwart a prevailing stream of moist air. The effectiveness of a mountain range in producing precipitation depends upon the angle it makes with the moisture bearing winds. Ridges at right angles have a marked relationship, but those parallel show little effect. The vertical lift imposed by the range is proportional to the sine of the angle of slope but the vertical velocity induced probably drops off exponentially with height. One suggested reduction is division of the vertical velocity by two every 1000 m.

The actual lift given by a mountain range is not normally very large, and on its own would produce only light rain, unless the range were very high. Frequently, however, the initial lift given by the range, releases latent instability of the air mass with resulting strong vertical motion and heavy rain. This 'trigger' effect or orographical intensification depends on the stability of the windward air mass, and can be estimated by comparing rainfall at adjacent sites in the valley and on the slopes. It would be expected from the drop off of precipitable water with height that rainfall would decrease with elevation, however, the deformed air field over a range gives improved conditions for the receipt of rain by converging the trajectories of the falling rain drops into the rain-gauge. The permanence of the range also gives a large number of occasions when some rain will fall. In fact, the main feature about rainfall as a range is ascended, is not the higher intensities received but the longer periods of rainfall, which are often of light intensity. The highest ratios of 24-hr rainfall to the precipitable water in the air mass is thus found in areas where topography assists rainfall.

In mountainous regions the variations in rainfall make assessment particularly difficult especially because sites for rain-gauges are few because of inaccessibility and are often non representative. Every source of data must be used to achieve good estimates of rainfall. The revised isohyetal method is helpful in that mean annual rainfall patterns, runoff figures or information from plant types may enable appropriate weights to be allotted to the rain-gauges. In rugged areas, precipitation depths should always be compared with runoff figures for corresponding periods to learn how representative the the gauges are within the area. As part of this study, information on soil, infiltration rates and runoff characteristics are also needed.

Objectively, rainfall may be related to the various parameters by either statistical or graphical coaxial correlation. Some of the variables normally encountered are (a) elevation, (b) slope or rise, (c) orientation or direction to which the gauge is open, (d) degree of shelter, (e) distance from the source of moist air, (f) direction and properties of the rain producing air masses, (g) meteorological or climatological region, etc. Some of the factors are mutually exclusive or are important in some areas but not in others. By testing these and any other locally significant parameters in the data for a region, the relevant factors may be found and the other disregarded. Spreën successfully developed a method for the drawing of normal isohyetal maps in a mountainous area of West Colorado by using some of these parameters and the technique of graphical, coaxial correlation. [12] Elevation, rise, orientation exposure, and zone were found to account for 85% of the precipitation while elevation alone accounted for 30%. It follows that if a reasonable correlation is established, the rainfall may be estimated in those areas not represented by rain-gauges.

Spillover, or the carrying over of the rain-drops or snow flakes formed above one catchment into a neighbouring one, can be assessed by a careful study of the air flow patterns over the range and the trajectories of the raindrops or snow flakes. The solution is peculiar to the valley studied, and is always an estimate.

11.15 The station-year method of extending records. [13] The records for a station may be extended by the station-year method. This involves using all records for adjacent stations or more precisely, in the one meteorologically homogeneous zone, and combining them as representing the record at one station for the total number of years used. For example, 10 stations each of 30 yr of record give 300 station-years of record. It follows that if the station-years are to represent a valid statistical series, each event must be independent of the others.

The two conditions of meteorological homogeneity and statistical independence appear to be mutually exclusive and may conflict. This means that the method should not be used unless both factors have been considered and are satisfied by the data.

It is implied that the data from all stations used possess similar frequency distributions. Thus, when all stations are used, the frequency distribution typical of the zone is more fully defined. It has been shown however, [14] that the use of a number of stations of varied years of record in the station-year method, essentially gives a weighted average of the moments of the distribution. The resulting probabilities are therefore only an average, weighted by the number of years at the stations used. The length of record is not extended by the method and most events are contemporaneous. However, the situation is different when the storm events at the stations are independent. For example, a local small area storm may, in its random path or life, give an isolated heavy fall at one station only and at no others in the zone. This fall may become a new member to the station-year statistical series which would not otherwise be included because of the location of gauges. Also, because of the random nature of the storm, it may be a comparatively rare event and so might be assigned a higher recurrence interval for the zone than for an individual station. The extrapolation of the record in this way should not give undue confidence in the accuracy of the method and must always be used with reservations.

The usefulness of the station-year method depends upon how representative the periods used are of the long term population. Usually the records are contemporaneous for most of the period covered, although it would be preferable for the station to encompass a wide spread of years. The occurrence of extremely wet or dry spells within the period of record may effect all stations and give a false idea of the true frequency distribution. For this reason, it is as well to assess the relation of the rainfall in the periods used to the longest available records.

In mountainous areas there are usually a multiplicity of meteorological factors which initiate, intensify, or inhibit rainfall. The rainfall data in such regions, even for stations close together, do not display the same frequency distribution. Accordingly, the station-year method is not used in mountainous country but restricted to localities where the meteorological factors are similar and consistent.

The definition of meteorological homogeneity used here is more restrictive than that employed later in the estimation of probable maximum precipitation and discussed in paragraph 17.4. The latter is concerned more with the possibility of occurrence of a storm type within an area, rather than the statistical distribution of the rain falls therein.

It may be concluded that in zones which are meteorologically homogeneous from the statistical point of view, and without marked topography, the station method can be used with reservations. This means recognising that the method is essentially one of averaging the frequency distribution over the zone to get an improved description of the statistical population and, when a rare event occurs at one station, to use this, as an indication only, of a probable return period. The recurrence interval obtained must be used with caution and not assumed to be a valid extrapolation of the period of record.

REFERENCES

[1] United States Weather Bureau (1957–8). Rainfall Intensity – Frequency Regime Pt 1 & Pt 11 *Tech. Paper No. 29, U.S.W.B.* Washington.

[1a] United States Weather Bureau (1961). Rainfall – frequency Atlas of the U.S. *Tech. Paper No. 40,* U.S.W.B.

[2] Miami Conservancy District (Rev. 1936). Storm rainfall of eastern United States. *Tech. Rep.* pt V, p. 258–9. Miami Conservancy District, Dayton.

[3] HUFF F.A. and NEILL J.C., (1957). *Illinois State Water Survey Bulletin,* No. 44.

[4] RICHARDS B.D., (1944). *Flood Estimation and Control,* Chapman and Hall, London.

[5] World Meteorological Organisation (1969). Manual for depth-duration area analysis of storm precipitation. *W.M.O. No. 237 TP 129.*

[6] U.S. Army Corps of Engineers. *Storm Rainfall in the United States* (to date) Depth-Area-Duration Data.

[7] WIESNER C.J., (1960). Analysis of Australian storms for depth-duration-area data. *Rain Seminar, Sydney. Com. Bur. of Met.,* Melbourne.

[8] PAULHAUS J. and KOHLER M.A., (1952). Interpolation of missing precipitation records. *Monthly Whr. Rev.* **80**, 129–133.

[9] McDONALD J.E., (1957). Note on precision of estimation of missing precipitation data. *Trans. Am. Geophys. Uni.* **38**, 657–661.

[10] KOHLER M.A., (1949). Double mass analysis for testing the consistency of records and for making required adjustments. *Bull. Am. Met. Soc.* **30**, 188–9.

[11] WEISS L.L. and WILSON W.T., (1953). Evaluation of significance of slope changes in double mass curves. *Trans. Am. Geophys. Uni.* **34**, 893–896.

[12] SPREEN W.C., (1947). A determination of the effect of topography on rainfall. *Trans. Am. Geophys. Uni.* **28**, 285–290.

[13] HAFSTAD K.C., (1942). Reliability of station-year rainfall frequency determinations. *Trans. A.S.C.E.* **107**, 633–683.

[14] FOSTER E.E., (1949). *Rainfall and Runoff,* Macmillan, London.

CHAPTER TWELVE

Snow

12.1 Definition and importance of snow. Snow was defined in paragraph 10.1 as a solid form of precipitation formed by the condensation of water vapour at temperatures below freezing point. Although originally formed as crystals, it is readily modified and compacted. Snow on the ground is a complex mass of flakes, crystals, air and water, and can possess densities varying from an average of 0·10 for freshly fallen snow to 0·91 for glacial ice. Although it may be measured in terms of depth and density, it is best assessed by its water equivalent.

In those catchments where snow falls it represents an important contribution to water storage within the basin. It is normally stored in the winter and becomes available as snow melt in the spring. Although it is of greater value for plant growth in spring, it is often in excess of requirements to produce flood hazards which demand control measures. Snow cover and associated low soil temperatures in the colder months affect runoff conditions to give high coefficients of runoff.

In probable maximum precipitation studies it may be found that the probable maximum summer rain storm may produce a smaller flood than a smaller spring storm combined with low infiltration, higher runoff, and snow melt in the spring. Therefore, in basins where sizeable snow accumulations may occur, the probable maximum precipitation is estimated on a seasonal basis to combine with probable snow melt.

Snow affects and modifies agricultural practices, the keeping of livestock, and the conservation of wild life. It has important recreational aspects. It influences transport, buildings, transmission lines, and water movement to the extent that special provisions must be made for it and development is restricted in these areas where it occurs frequently.

12.2 The measurement of snow. Snow is measured by a gauge similar to the common rain-gauge but of greater capacity and without a funnel on which the snow would collect. The snow is melted and its water equivalent recorded. Storage snow gauges are used for measurement in remote areas and read at convenient intervals. Weighting methods are often used with antifreeze and oil used as protection against freezing and evaporation. Due to the low density of the snow flakes, they are easily carried by wind out of containers or catchments. Turbulence over the orifice of the gauge can be minimised by careful siting or by the use of shields such as the Nipher, Alter, or Shasta types, (see paragraph 10.3). Some types of pluviographs operated in regions affected by frost and snow are heated to prevent freezing.

Snow on the ground can be measured by graduated sticks or permanent snow stakes which are calibrated and carefully located. The amount of new snow can be determined from snow boards which are placed on old snow in sites where the accumulation pattern is known to be representative of the region. The depth above the board is a measure of the new snow.

The erratic and irregular movement of snow during a storm makes its measurement by snow gauges an inaccurate process and it is better measured by depth over a catchment. Depth of snow alone does not give the water equivalent which equals snow depth times snow density.

$$\text{Density of snow} = \frac{\text{volume or depth of water equivalent}}{\text{volume or depth of snow}}. \qquad (12.1)$$

Snow density varies from 0·004 to 0·91 with an average value of 0·10. Both factors are considered in snow surveys which take and measure samples by coring and weighing of the snow pack along an established line or snow course. The snow course is carefully selected to give readings representative of the catchment. The best sites are found from tests and experience within the region.

Snow gauges using radio-isotopes have been developed [1,2] and, in association with telemetering equipment, enable the recording of snow and water equivalents in remote areas. The results are comparable with those given by snow surveys.

12.3 The rate of snow melt. The amount of snow on a catchment represents additional water storage which may be made available as runoff when melting occurs. The rate of snow melt plus the precipitation rate and appropriate runoff relationships are necessary to predict stream flow. Rainfall and snow melt can proceed separately or simultaneously and both must always be taken into account.

Precipitation rates have already been discussed and snow melt rates must now be considered. One approach is to examine the physics of snow melt and use a series of empirical equations. Constants for these equations have been derived, mainly in the United States, but these have a wider application. Another method relates the snow melt rate to air temperature as the principal variable, and then determines the constants of the equation from available data. Snow melt curves giving rates or accumulated amounts in terms of a temperature index, usually also accumulated, are constructed for the valley in question.

12.4 The physics of snow melt. In the study of evaporation the energy balance approach is used to obtain the energy necessary to convert water into vapour. A similar energy balance can be set up for a snow surface and from a knowledge of the energy available and necessary to convert ice or snow into liquid, the amount of snow melt is determined.

The principal sources of energy for snow melt are

(*a*) Net radiation, R_n, discussed in chapter 2.

(*b*) Heat from the condensation of vapour.

(*c*) Heat of conduction from the overlying air, ground, or rainfall.

Each of the above factors may result in either a gain or loss of energy to the snow surface and are inter-related. Equations may be set up involving them to give the amount of snow melt. These equations estimate the theoretical melt from a flat snow surface. Data from particular valleys are necessary to give the empirical constants and basin factors.

12.5 Snow melt from radiation. The relationship between net radiation and snow melt is comparatively simple, about 200 cal/cm²/day resulting in an inch of snow melt. The albedo or reflectibility of the snow surface is significant, and its high value which goes up to 0·9, reduces the solar radiation penetrating the surface. Together with back radiation this explains why snow melt does not always take place or proceeds slowly in clear weather. Use of a net radiometer and information on albedo in a representative site is a method of evaluating snow melt due to this source. When radiometers are not available, the methods of chapter 2 can be used to estimate the net radiation. Alternatively, simple equations relating snow melt to cloudiness using time of year, average albedoes, latitude, and tables of radiation can be used. For example, the 12-hr, midday–midnight, snow melt M is related to cloudiness, C, 0 to 1, and to M_0, the 12-hr melt, in clear weather between lat 40 to 48 degrees by the equation

$$M = M_0 \left(1 - \tfrac{3}{4} C\right). \quad (12.2)$$

Table 12.1 gives values of M_0 for months when melt occurs in latitude 40–48 and with various types of snow surface.

Table 12.1. *Snow melt (M_0) from midday–midnight, lat. 40–48 degrees and clear weather.*

Month		Snow melt (12hr)		
N. hemisphere	S. hemisphere	Average	Fresh snow decrease 20%	Old snow increase 20%
March	September	0·35	0·28	0·42
April	October	0·42	0·34	0·50
May	November	0·48	0·38	0·57
June	December	0·53	0·42	0·63

For other latitudes, tables or diagrams of solar and terrestrial radiation may be used to find the relative factors.

12.6 Snow melt from vapour condensation. The heat from the condensation of water vapour causes snow melt proportional to the wind velocity and the vapour pressure gradient. The 6-hr snow melt M in in. can be expressed in terms of the average wind, v, at 50 ft in miles/hr and the vapour pressure of the air e_d in mb, measured at 10 ft by the equation

$$M = 0{\cdot}00578v\,(e_d - 6{\cdot}11). \quad (12.3)$$

12.7 Snow melt from conduction and convection. The heat reaching the snow surface by convection and conduction from the air is proportional to the wind velocity and the temperature gradient. Allowance must be made for the variation of temperature with height. The 6-hr snow melt in in.

$$M = 0{\cdot}000184 \,.\, 10^{-0{\cdot}0000156\,z}\, v \,.\, (T_a - 32) \quad (12.4)$$

where z is the height above MSL in ft, T_a is air temperature in °F at 10 ft and v miles/hr at 50 ft.

If heat reaches the snow surface by conduction from the ground the snow melt in in.

$$M = 1{\cdot}29\ (T_s - 32)\sqrt{t} \qquad (12.5)$$

where T_s is soil temperature and t is the time in hr of persistence of the temperature gradient. In most cases T_s is near 32 at the critical times considered and this factor is therefore zero.

The snow melt caused by rain of P in. is

$$M = \frac{P\,(T_w - 32)}{144} \qquad (12.6)$$

where T_w is the wet bulb temperature at 10 ft in °F.

12.8 Theoretical estimation of snow melt. The formulae 12.2 to 12.6 can be applied, when relevant, either singly or in combination. The theoretical snow melt applies to a smooth exposed surface, wind velocities measured at 50 ft, and wet and dry bulb temperatures measured at 10 ft above the snow surface. A logarithmic wind profile can be used to convert winds at other heights and a lapse rate of 3·5°F per 1000 ft to adjust temperatures. Alternatively, a 1/6th power profile may be assumed and the height adjusted to 10 ft and 50 ft for temperature and wind by multiplying respectively by $1{\cdot}47z^{-1/6}$ and $1{\cdot}92z^{-1/6}$ where z is the height of measurement in ft.

The formulae or, combinations of them, may be expressed in graphical form to simplify the estimation of snow melt. [3–5]

In a natural basin there are many variations in slope, aspect, forest cover, albedo, etc. and actual snow melt will accordingly differ from the theoretical amount. A basin factor equal to actual snow melt divided by theoretical snow melt can be determined for the basin from past data and applied on future occasions. The basin factor should theoretically be 1 for a smooth, exposed, snow surface but in most catchments has a value closer to 0·5.

The United States Army Corps of Engineers [6] has used the theoretical relationship and extensive data to obtain equations for snow melt during rain and in fine weather. These equations take into account cover and topographical factors and are adaptable to most situations.

12.9 Estimating snow melt by correlation methods. Theoretical estimations of snow melt are not always successful because of the many variations in the meteorological parameters over a basin, the difficulty in measuring them, and their complex inter-relations. Empirical relationships between the variables may be established for particular valleys by using the above theory, past data on snow melt, and the meteorological records.

The correlation methods of statistics or the coaxial graphical correlation techniques can be used to relate snow melt to one or more of the variables mentioned in the physics of snow melt. A good fit to past data should be sought. The relationship so established may be used to estimate future snow melt.

Air temperature is related to radiation, wind velocity and humidity and is the most important single variable in the snow melt equation. Accordingly air temperature, or a derived index involving it, can be empirically correlated with snow melt. The diurnal range or difference between daily maximum and minimum temperatures, is one index. Another is the number of degree days above an accepted temperature which is commonly the freezing point. For example, a mean daily temperature of 42°F represents 10 degree days above 32°F, 12 degree days above 30°F or 2 degree days above 40°F, depending on the base found convenient for computing melt. The most suitable base temperature is decided from the nature of the observation site and the valley. Degree days can be determined accurately from the hourly trend of temperature but is not always available.

Typical correlation equations established in the United States and giving springtime daily melt in inches are,

for open sites, $M = 0{\cdot}06\ (\bar{T} - 24)$ or $M = 0{\cdot}04\ (T_{\max} - 27)$ (12.7)

for forest sites, $M = 0{\cdot}05\ (\bar{T} - 32)$ or $M = 0{\cdot}04\ (T_{\max} - 42)$. (12.8)

These are applicable for mean temperatures from 36–66°F or for maximum temperatures from 44–76°F.

At a point within a catchment the rate of snow melt per degree day varies from 0·015 to 0·20 in./degree day above 32°F with an average of 0·05 in. For the whole catchment a basin factor is used and values between 0·02 and 0·11 with a maximum of 0·30/degree day above 32°F have been determined. More accurate equations can be determined for particular valleys from the data in that valley and using other variables.

Snow melt is usually translated into runoff by using appropriate rainfall–runoff relationships. However, the connection between degree days and runoff may be determined directly for particular valleys. If the accumulated runoff is plotted against accumulated degree days, an S-shaped curve is obtained for the valley in question. The shape is due to the close association between snow melt, subsurface flow and the stream recession. These relationships are particularly useful if based on long hydrological records and permit rapid computation of runoff.

12.10 The extent and place of occurrence of snow melt. It is necessary to know both the extent of snow melt and where it occurs in a basin. The accumulation of snow can be reasonably well defined by its elevation or the snow line. Also, the area or proportion of a valley covered by snow at any time can be expressed in terms of the contours of the snow line. As melting proceeds, however, snow does not disappear or retreat along contours but according to its slope, aspect, and degree of cover. That is, snow in a cleared area melts more rapidly than in forest land and slopes, directly inclined to the sun's radiation or wind are freed of snow sooner than those not so favourably placed to the melting agents. The melting snow line is expressed as an average contour. No melt occurs as long as the freezing isotherm is below the snow line, but, as it rises above the snow line, melt occurs in the area between it and the snow line. The extent of this area of melt can be determined by using the lapse rate of temperature, 3½°F per 1000 ft, the temperature of the index station, the average elevation of the snow line, and the area–elevation relationship for the catchment. The percentage or proportion of the catchment under melt determined in this way multiplied by degree days gives a weighted value of degree days for the catchment. The snow melt for the basin can then be determined from the product of the weighted degree days and basin melt factors. The difficulty in evaluating the various factors, however, gives a lower level of accuracy and reliability than might be expected from the physical equations.

REFERENCES

[1] ITAGAKI K., (1959). An improved radio snow gauge for practical use. *J. Geophys. Res.* **64**, 375–383.

[2] ROBINSON C., (1960). Gamma radiation gauges snow pack water content. *Elec. World,* **154**, 82–83.

[3] BERNARD M. and WILSON W.T., (1941). A new technique for the determination of heat necessary to melt snow. *Trans. Am. Geophys. Union.* **22**, 178–181.

[4] WILSON W.T., (1941). An outline of the thermodynamics of snow melt. *Trans. Am. Geophys. Union.* **22**, 182–195.

[5] LIGHT P., (1941). Analysis of high rates of snow melting. *Trans. Am. Geophys. Union.* **22**, 195–205.

[6] U.S. Army Corps of Engineers (1960). Runoff from snow melt. *U.S. Army Engineering Design Manuals.* E.M. 1110-2-1406.

OTHER REFERENCES

CHOW V.T., (1964). *Handbook of Applied Hydrology,* McGraw Hill, New York.

BUTLER S.S., (1957). *Engineering Hydrology,* Prentice Hall, New Jersey.

CHAPTER THIRTEEN

Water available for precipitation

13.1 Precipitable water within an air mass. The water content of an air mass is conveniently expressed in terms of its precipitable water, W. This is defined as the equivalent linear depth of water within a column of air if all the water vapour is condensed over the base area of the column. There is no known meteorological process which does this and the term is therefore not a good one. More precise and preferred expressions are the liquid equivalent of the air mass or the water equivalent of the air mass. However, as these are not in current use, the commonly accepted term of precipitable water is used here.

13.2 Computing the precipitable water. [1] Aircraft flights, radiosondes and other instruments give measurements of the vertical structure of the atmosphere, usually in terms of pressure p in mb, temperature T in °C and specific humidity, q, in g/kg or equivalents.

Specific humidity $q = \dfrac{\text{density of water vapour}}{\text{density of air}} = \dfrac{\rho_w}{\rho}$ and is in g/kg.

The mass of water vapour in unit column

$$dz = \int_0^z \rho_w \, dz$$

$$= -\int_{p_0}^{p_z} \frac{\rho_w}{\rho} \frac{dp}{g} \quad \text{using equation (7.1) } dp = -\rho\, g dz$$

$$= -\frac{1}{g} \int_{p_0}^{p_z} q \, dp.$$

Express this as depth, W, using p in mb and q in g/kg.
We get

$$W \text{ (in.)} = 0{\cdot}0004 \int_{p_z}^{p_0} q \, dp, \qquad (13.1)$$

or

$$W \text{ (in mm)} = 0{\cdot}01 \int_{p_z}^{p_0} q \, dp.$$

This equation is used to determine the precipitable water between two levels in an air mass from information on the variation of humidity and pressure with height. For example, if humidity, q, varied with height, z, or pressure, p, as shown in Fig. 13.1, then the precipitable water between two levels is the area bounded by the curve and the appropriate pressure levels. That is, precipitable water between p_1 and p_2 is

$$W_{12} = 0{\cdot}0004 \int_{p_1}^{p_2} q\, dp = \text{area } A.$$

Also precipitable water between p_3 and p_4 is

$$W_{34} = 0{\cdot}0004 \int_{p_3}^{p_4} q\, dp = \text{area } B.$$

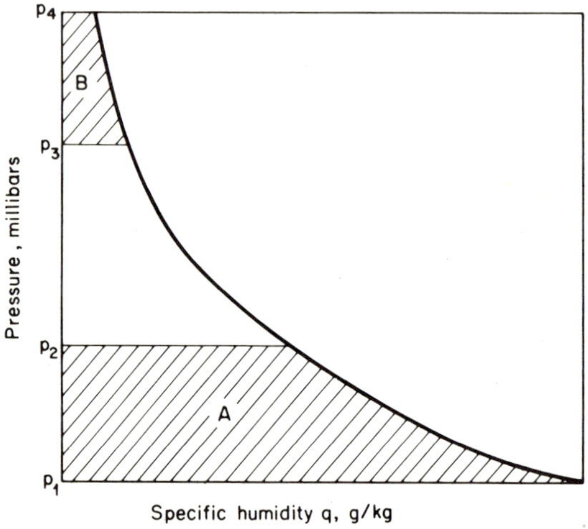

Fig. 13.1. Precipitable water from p–q relationship.

13.3 Precipitable water from aerological observations. If the variations of specific humidity with height or pressure are known from radiosonde or other observations, the precipitable water may be estimated as in Table 13.1.

13.4 Precipitable water from surface data. Vertical soundings of the atmosphere have been available only recently and do not always fully represent the air mass over an area nor throughout the duration of a storm. Surface dew point observations are more numerous and their variations with time and area can be determined. In storm situations with rain, it is assumed that the air mass is saturated and, the vertical humidity distribution is represented by the surface dew point, decreasing with height at the saturated pseudo adiabatic lapse rate. With these assumptions, equation 13.1 can be used to obtain values of the precipitable water from the surface dew points to the required heights.

The precipitable water within the layers of a saturated air mass therefore can be expressed in terms of its surface or 1000 mb dew point and the respective pressure levels. Table 13.2 gives, in hundreds of inches, the precipitable water between the surface and the levels indicated for air masses of certain surface dew points with assumed saturated pseudo adiabatic lapse rates.

Table 13.1. *Calculation of precipitable water (W) from air mass sounding*

Level n	Pressure p mb	Humidity q g/kg	Pressure difference $p_n - p_{n+1}$	Twice mean q $2\bar{q}$ $q_n + q_{n+1}$	Product (4) × (5)
1	1005	14·2			
2	850	12·4	155	26·6	4123
3	750	9·5	100	21·9	2190
4	700	7·0	50	16·5	825
5	620	6·3	80	13·3	1064
6	600	5·6	20	11·9	238
7	500	3·8	100	9·4	940
8	400	1·7	100	5·5	550
9	250	0·2	150	1·9	285
					10215 × 0·0002 = 2·04 in.

Table 13.2. *Depth of precipitable water in hundredths of inches between* 1000 mb *and indicated height.*

Height (ft × 100)	1000 mb 60	65	Dew point 70	75	78
2	03	04	04	05	06
4	06	07	09	10	11
6	09	11	13	15	17
8	12	15	17	20	22
10	15	18	21	25	28
15	22	27	31	37	41
20	29	35	41	48	53
25	36	43	50	60	66
30	42	50	59	70	78
40	53	64	76	90	100
50	64	77	92	109	121
70	82	99	119	143	159
100	102	125	152	184	206
150	123	153	189	232	262
200	133	168	210	262	297
300	138	177	225	285	329
400	138	177	227	289	335

The precipitable water in the saturated column of air above 1000 mb is also shown in terms of the surface dew point and the pressure level or height of the column by Fig. 13.2, a and b. Here, pressure is the ordinate and precipitable water, the abscissa. The curved lines represent 1000 mb, or surface, dew points and the abscissae of their intersections with any pressure levels indicate the precipitable water from the surface to that level for the particular dew point. For example, saturated air with surface dew point 70°F contains 1·5 in. of precipitable water between 1000 mb and 700 mb, and 2·3 in. between 1000 mb

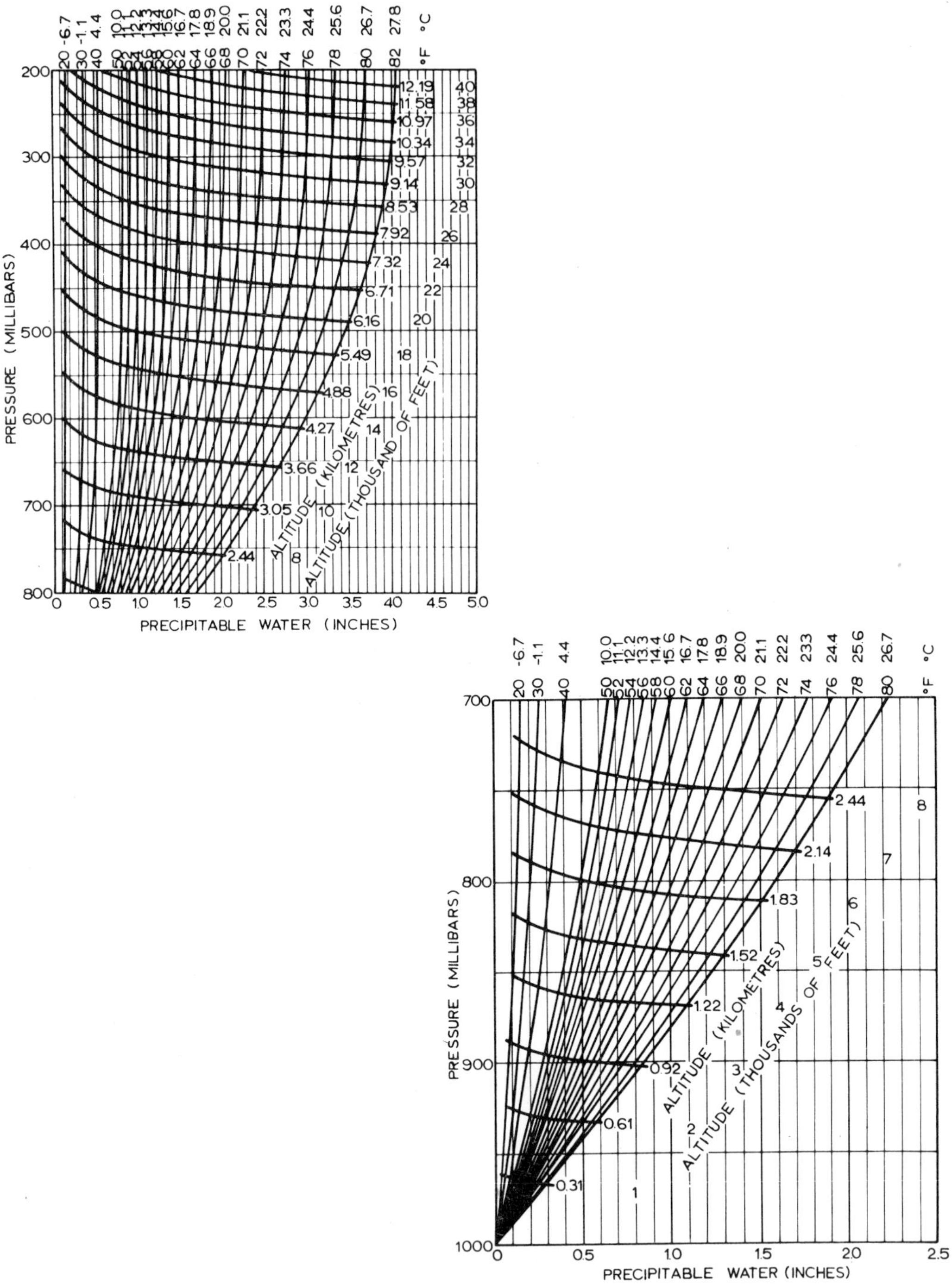

Fig. 13.2. Precipitable water for saturated air mass knowing surface dew point.

and 200 mb. It follows that there is 2·3 − 1·5 = 0·8 in between 700 mb and 200 mb. Instead of pressure, a height scale shown on the right and represented by a series of concave curves may be used. Figure 13.2 is a convenient and quick method of obtaining the precipitable water in a saturated air mass from the 1000 mb dew point to a given height or between known pressure levels. It is used in storm studies and the estimation of probable maximum precipitation. If greater accuracy is required tables or nomograms of precipitable water are used. [2–4]

13.5 Comparison of methods. The analysis of a number of storms in the United States represented by aerological as well as surface data showed that precipitable water from observed vertical data, W_{obs}, equals 0·02 plus 0·99 times the precipitable water estimated from the surface dew point, W_{est}.

i.e.
$$W_{obs} = 0{\cdot}02 + 0{\cdot}99\, W_{est}. \qquad (13.2)$$

The correlation coefficient was 0·92 and the standard error of estimation 0·07 in. [5] Additional storms over more extensive areas also show good agreement between the two methods. It may be concluded therefore that in storm situations, the use of surface dew points, carefully observed, preferably at a number of locations and over a reasonable period of time, gives satisfactory estimates of precipitable water under conditions leading to rainfall. Surface observations of dew points have the advantage that they are more numerous, have been measured for a considerably longer period, and their time and areal variations are known or can be assessed.

13.6 Condensed moisture held in an air mass. The precipitable water of an air mass is the liquid equivalent depth of the water vapour within that air mass. If condensation occurs with the appearance of cloud, fog, or other precipitation forms, and moisture inflow continues to replace that condensed and to maintain the same moisture stratification, then the condensed products in liquid or solid form represent an additional quantity of water held over the area, usually by vertical currents. This may even equal or exceed the precipitable water which is held as vapour. This stored water can give high rates of precipitation for short periods, of the order of minutes, when released from the air mass say, by the failure of the vertical holding currents. Cloud burst rainfalls are examples. Strictly speaking, in precipitation studies, stored water in an air mass should be evaluated both before and after precipitation. It represents a storage term in the overall moisture balance of the storm system.

In practice this is neglected for a number of reasons. Firstly the intense rain which results from the sudden release of the stored water occurs over a small area and for a short duration as it is soon exhausted. Runoff which occurs may be significant in flows over roofs and in gutters but not on the larger catchments considered in hydrology. Usually intense rains from storms over sizable areas, say in excess of 100 square miles and of longer durations, more than 1 hr, are significant. Again, there are difficulties in computing and measuring the stored water above an area. Furthermore, because cloud is present both before the start and after the end of precipitation, there may be only a small difference between the water stored at the beginning and end of a rain burst. Thus, at the commencement of the rain burst the air above the catchment contains precipitable water in the form of vapour plus condensed water in the form of cloud. During precipitation there is a continued flow of moisture and its release but at the end of the rain burst, the amount of water held by the air mass as vapour and cloud may not be vastly different from that at the beginning of the rain burst. The significant factor contributing to rainfall intensity is the rate of supply of moisture during the time period. This is measured by the precipitable water, W, times the inflow velocity, both evaluated with respect to time,

i.e. $$\int_0^t Wv\, dt.$$

13.7 Representative surface dew points. If surface dew points are to indicate the moisture content of air masses during storms, the greatest care must be taken to assure that they represent the air masses over periods of time and in space, both over areas and in the vertical. Basic data must be carefully checked to assure that instruments were operating satisfactorily at the times considered. For example, thermographs and hygrographs must be correctly calibrated and the wet bulb supplied with clean and adequate water. A check should be made that abnormal and unrepresentative conditions, such as stagnant air with high local humidities, were not being experienced at the time. The surface dew points should indicate that the air mass maintained its properties over a wide area and persisted for a reasonable duration which may be related to the storm duration. During the storm one can expect the air mass to be saturated throughout its depth and well mixed.

In order to satisfy these conditions, all available dew point stations are used within and beyond the storm area and particularly towards the warmer moist air along the likely trajectory of the air mass. To give a common reference, the dew points at different elevations are compared by reducing them to the standard level of 1000 mb. This is done using the saturated pseudo adiabatic lapse rate. Figure 13.3 shows a simplified chart which can be used for this dew point adjustment. In this diagram a point, corresponding to the elevation given as ordinate, and the observed dew point given as abscissa, is taken and the saturated pseudo adiabat, or the sloping curved line, is followed down to the intersection with the base line from which the 1000 mb dew point is read off. This procedure shows up anomalies which can be investigated and corrected or eliminated. The consistent dew points are averaged to become the representative dew point of the storm. The areal variation of the dew points is shown by plotting 1000 mb dew points for all stations near the storm on a map. Isopleths should show uniformity and consistency. Normally, storms of some hours duration and covering areas exceeding 100 square miles are considered so a broad flow of warm moist air persists over an appropriate period of time. For a storm of D hr duration, it is possible to select the highest D hr persisting dew point or the highest dew point that can persist for D consecutive hours. For reasons which appear later, it is the practice to select the highest persisting dew point for a duration of 12 or 24 hr.

Continuous or hourly dew point readings permit an examination of the variations of the high dew points with duration. The highest persisting dew points for 1, 3, 6, 12, 24, 72, etc. hr can thus be determined, and if storm models are being used, give the appropriate precipitable water for the duration in question. When actual storms are adjusted by the use of ratios of precipitable water, it is sufficient to take the highest persisting dew point for a duration appropriate to the size and duration of the storm, provided the same

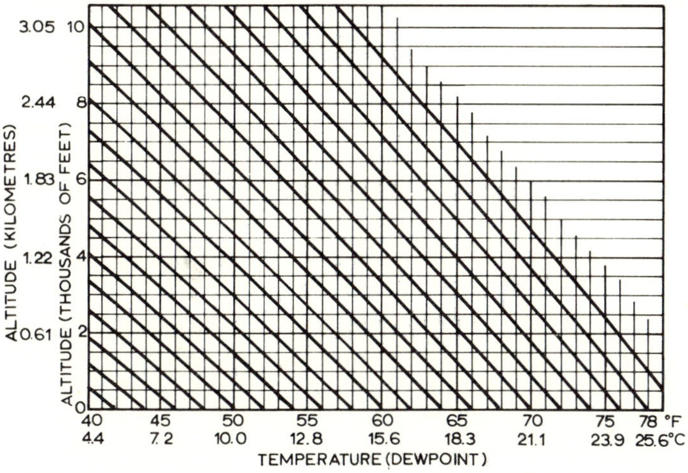

Fig. 13.3. Pseudo adiabatic chart for adjusting dew point with height.

duration is used for the maximum dew point. As a general rule however, the highest dew point that persists for 12 or 24 hr is considered satisfactory and representative of the air mass flowing into a storm. The same duration is also used for the maximum dew point.

There are many peak and short term variations in the water content of air masses flowing into storms, and these are associated with short intense bursts of rainfall especially over small areas and for short durations. However, observational techniques and theories are not precise enough to fully assess such effects and variations. Thus the longer duration dew points are preferred for adjustment of actual storms. Also, these variations make it difficult to use storm models for small areas and short durations, although storm models are successful on the larger scale.

13.8 The variation of dew point with duration. In a number of storm situations the variation of persisting dew points with duration has been examined. [6] The dew point decreases as the duration increases with a fall of about 5°F from 1 hr to 24 hr and a further 5°F fall from 24 hr to 96 hr.

By plotting the precipitable water corresponding to the highest persisting dew point value for various intervals up to 72 hr and drawing smooth decay curves, it is possible to express the precipitable water at one duration in terms of that at another. [7] This is shown in Table 13.3 where the precipitable waters at durations between 1 and 72 hr are expressed as a percentage of the precipitable water at 12 hr.

Table 13.3. *Variation of moisture in storm with duration.*

Duration (hr)	1	3	6	12	18	24	36	48	72
% of precipitable water at 12 hr	107	106	104	100	97	95	91	88	83

In the United States there is no significant geographical variation in the moisture decay curves. Seasonal variations are about 2 or 3 percent and are considered of slight significance only. Table 13.3 is used to extend 12 hr or 24 hr persisting dew points to other durations in the absence of observations covering these durations.

13.9 Determining the surface dew point representative of a storm or a locality. In storm studies it is necessary to estimate dew points which are representative of the air flowing into the storm and the maximum values which that air mass might achieve in the same or adjacent areas. The trajectory of the air involved is significant and detailed synoptic analyses are often necessary to determine the path of the air mass, its water content, the moisture barriers involved and the changes likely to be made to its properties. Knowledge of the source region enables logical extrapolations or transpositions to be made.

In evaluating the representative surface dew point of a storm it is necessary to take all precautions mentioned in paragraph 13.7 and use either the 12 hr or 24 hr persisting dew point. Where continuous or hourly records of dew point are available, these are used to determine the highest 12 or 24 hr persisting dew point. This should not be higher than the minimum temperature within the period, which is used as a check.

If dew point measurements are made at set times, the highest persisting dew point is taken as the lowest of these readings, and the intervening minimum temperature. When readings are made at 9 a.m. only, the 24 hr persisting dew point only can be determined and is the lowest of the two adjacent 9 a.m. readings, and the intervening minimum. Conversion to other durations is approximated by Table 13.3. Due to deficiencies in the reporting networks of some countries, the 24-hr persisting dew point is used by necessity.

13.10 Maximum values of surface dew points. To determine the probable maximum precipitation, values of the maximum persisting dew points are required. When the P.M.P. is to be estimated on a seasonal basis, maximum dew points are determined for each month of the year. Several methods are available, and where possible all should be used and compared.

In the first method, all available and checked dew points in the region for the selected duration, 12 hr or 24 hr, are reduced at the saturated pseudo-adiabatic lapse rate to 1000 mb and plotted against the time of year when they occurred. A smooth enveloping curve, encompassing all values, is drawn to give the seasonal maximum observed dew points. If higher values occur in subsequent months, the curve is correspondingly modified. Due to desiccation by the passage of the air masses over land and mixing to higher levels, the highest 1000 mb dew points are usually encountered at the low level stations near the coast.

The highest dew points for each month over the length of record may also be fitted to a statistical distribution and the once-in-five, twenty-five or one-hundred-yr dew point for that month estimated. Among the distributions which have been used are the Fisher–Tippett Type 1 or Gumbel, the normal, and the log–normal. Each of these are most expeditiously handled by using the appropriate probability papers. Seasonal curves for each of the recurrence intervals are drawn and compared with the envelope of maximum values.

As the rain giving air mass in a storm situation originated from an ocean where it acquired the properties of the underlying water surface, the seasonal mean sea temperatures of the source region is related to the maximum dew point which the air mass might attain.

When the sea temperature is greater than the dew point of the air mass in contact with it, there is an upward flow of moisture. If however, the dew point exceeds the sea water temperature, condensation of water on the surface is expected. The dew point is normally less than the sea surface temperature as witnessed by the vapour pressure gradient from sea to air. Convection and horizontal movement or turbulence of the air mass distributes the moisture in depth and area while some moisture is stored in cloud layers or lost by mixing with drier air of higher latitudes. However, no firm relationship between sea temperatures and surface dew points can be given and differences of from 1 to 4°F have been determined. One of the difficulties in accepting sea temperature data is that much of it is taken in circumstances different from storm situations. The mean surface temperatures for the respective months of the year for the known source regions, however, indicate the maxima which the persisting dew points might reach. Thus where reliable sea temperature data is available it can indicate the likely maximum dew points of an air mass originating in the locality.

Sea water temperatures of 82°F and higher and storms involving air masses with surface dew points approaching 78°F indicate that a likely maximum dew point over the land at 1000 mb is 78°F for a period of 12 hr. Until further evidence is produced, this can be considered as the maximum 1000 mb dew point for an air mass extensive enough to feed into a storm of significant size and duration at least 100 square miles area and 1 hr duration. It is recognised that high sea temperatures and surface dew points occur especially in stagnant situations, over small areas, and for brief durations. In these cases the moisture is often concentrated in the surface layers and is not being fed into a storm. In using surface dew points in hydrometeorology to estimate precipitable water, the air is assumed saturated and near the pseudo-adiabatic lapse rate up to a considerable height. In storm situations, there is sufficient turbulence and mixing to guarantee that this occurs but in static situations such as in the centre of high pressure systems this is unlikely and dry high surface dew points are not admissible as being unrepresentative of the vertical structure.

REFERENCES

[1] SOLOT S.B., (1939). Computation of depth of precipitable water in a column of air. *Monthly Whr. Rev.* **67**, 100–103.

[2] U.S. Weather Bureau (1951). Tables of precipitable water and other factors for a saturated pseudo-adiabatic atmosphere. *Technical Paper No. 14* U.S.W.B.

[3] SHOWALTER A.K., (1954). Precipitable water Template. *Bull. Am. Met. Soc.* **35**, 129.

[4] PETERSON K.R., (1961). A precipitable water nomogram. *Bull. Am. Met. Soc.* **42**, 119–121.

[5] U.S. Weather Bureau (1960). Generalised estimates of probable maximum precipitation for the United States west of the 105th meridian for areas to 400 square miles and duration to 24 hr. *Technical Paper No. 38* U.S.W.B.

[6] U.S. Weather Bureau (1945). Revised report on maximum possible precipitation, Los Angeles area, California *Hydromet. Report No. 21B.*

[7] U.S. Weather Bureau (1961). Interim report. Probable maximum precipitation in California. *Hydrometeorological Report No. 36.* U.S.W.B.

OTHER REFERENCES

REITAN C.H., (1960). Distribution of precipitable water vapour over continental U.S. *Bull. Am. Met. Soc.* **41**, 79–87.

REITAN C.H., (1963). Surface dew point and water vapour aloft. *J. App. Met.* **2**, 776–779.

CHAPTER FOURTEEN

The processes and intensity of precipitation

14.1 Necessary conditions for the precipitation process. Before the moisture present in the air can be removed or precipitated the air mass must be subjected to a cycle of changes termed the precipitation process. In general, the air mass is 'processed' in a storm model between the time it flows into the model until it flows out and several stages can be resolved. In fact, before precipitation can occur and be recorded, the following conditions must be fulfilled.

(*a*) A sufficient supply of moisture, generally in the form of water vapour, must be available.

(*b*) The air containing the water vapour must be cooled to below its condensation point.

(*c*) The water vapour must aggregate or condense into water or ice particles.

(*d*) These particles must grow to such a size that they fall and reach the precipitation gauge.

The subject of cloud physics considers each of these stages in detail and a brief description only is given here.

14.2 The moisture in the atmosphere. Chapters 5 and 13 discuss the water contained in the atmosphere and its measurement. Usually water as a vapour is measured conveniently or more precisely as the liquid equivalent or the precipitable water, W. Moisture is also present in the atmosphere as a liquid and solid, and becomes visible in the form of cloud and precipitation products. The difference between the states can be expressed in terms of size with the water molecule in the vapour phase approximating to $10^{4}\mu$, cloud droplets 5–100μ, drizzle 100–500μ, and raindrops 500–5000μ. Snow, ice, and hail attain still larger sizes. At the one time, a cloud may contain water particles of a wide range of sizes and states, all of which are supported by vertical currents of varying velocities.

The precipitable water of an air mass does not include water in its liquid or solid form. It is the equivalent water depth of its water vapour content. However, it is observed that with fine weather and clear skies, the precipitable water is low but higher values are attained with cloudy conditions [1] and with rain, because the air is more fully charged with water vapour on these occasions. Because the saturation vapour pressure increases with temperature, so the precipitable water at a location attains higher values in the summer season. In general, the precipitable water at a place depends upon the depth of the air column above it, its latitude, altitude and distance from the sea, the overall atmospheric circulation, and the particular meteorological situation. An average annual value of precipitable water over a continent in middle latitudes is about 0·7 in., varying from a monthly average of 0·37 in. in winter to 1·2 in. in summer. On individual days it is as low as 0·1 in. or exceeds 3 in. [2,3]

Clouds may contain from 0·1 to 4 in., of equivalent water depth as water droplets and ice. This is additional to the vapour content of the air or precipitable water (W). As continuing rates of precipitation are the main interest, and cloud is present both before and after precipitation, although perhaps at different levels of water content, no large error is made by using W as the measure of water vapour content of the inflowing air in the computation of precipitation. For very short duration and small area storms, for example, the so called cloud bursts, the accumulated ice and water droplets in the cloud may add to the precipitation depth. Rough computations can be made from the work of cloud physicists but generally such estimates are not necessary and are neglected here.

14.3 Cooling of the air mass. At any given temperature there is a limit to the amount of water vapour which the air can hold. The air is saturated when it contains the maximum amount for that temperature. The variation of saturation vapour pressure with temperature is shown in Fig. 5.1. If an unsaturated air mass is cooled, its capacity to retain water vapour decreases, and as cooling continues, it reaches a stage where it is saturated or at its dew point. Condensation usually occurs at or near this point.

Masses of air are cooled by direct contact with surfaces at lower temperatures. Usually, if the air temperature is reduced below its dew point, condensation occurs and precipitation products become apparent. Dew, frost, mist, and fog occur in this way when moist air comes into contact with cold surfaces.

Layers of moist air also cool by radiating portion of their heat to become saturated with subsequent condensation products. Patches of radiation fog commonly form in this way. However, the most successful method of cooling large masses of air is by reducing pressure without loss or gain of heat, or by the process of adiabatic expansion. As pressure decreases with height, such a process is achieved by lifting the air mass adiabatically. This process is also known as dynamic cooling. Under natural conditions air masses are lifted and cooled in this way by the action of (*a*) convergence, (*b*) fronts, (*c*) topography, and (*d*) turbulence, or atmospheric instability. These processes occur singly or in combinations and merit further description beyond that already given.

(*a*) *Convergence* implies the movement and concentration of air into a particular area. Momentum acquired by air in other latitudes, the effect of low pressure areas or isallobaric convergence, the influence of topographical features, and the presence of divergence aloft may all cause air to converge within the lower layers with forced ascent and cooling. These processes are observed in the intertropical frontal zones, low pressure troughs, converging valleys with inflowing winds, and in tropical cyclones.

(*b*) *Frontal lifting* has already been mentioned in chapter 8. In a warm front, the warmer air rises over the cold with slopes of from 1 in 100 to 1 in 300, whereas, in the cold front, the cold air with a steeper slope of 1/25 to 1/100 cuts into the warmer air which is forced to rise. The steeper slopes in cold fronts produce the more rapid lifting and condensation with more intense rainfall but often of shorter duration.

(*c*) *Orographical lifting* occurs when air is forced to ascend a topographical barrier in its path. The air flow is distorted and the vertical velocity component is proportional to the land slope, being a function of the angle of slope and the horizontal wind velocity at right angles to the barrier. The vertical velocity decreases with height with

one suggested rate being a reduction by ½ every 1000 metres. [4] As land slopes can be steeper than frontal surfaces and permanent in location often the air is cooled more rapidly and for longer periods by orographical rather than by frontal lifting, other conditions being equal. This results in heavy rates and amounts of precipitation in mountainous regions. In all lifting processes the stability of the air is significant and must be taken into account. Orographical lifting commonly triggers off latent instability.
(*d*) *Turbulence* occurs on both a large and small scale, dependent on the size of the eddies. Small scale turbulence is set up by the movement of an air stream over physical obstructions with the creation of small eddies. The upward motion in these causes moisture to be carried up to the condensation level with the formation of shallow cloud. Large scale turbulence, convection or instability is related to the lapse rates of dry and moist air, the creation of large cells, and the lifting of air to considerable heights, at times to the tropopause. Instability is developed in an initially stable air mass in the following ways:

(*i*) *Heating from below* occurs when the surface is strongly heated as on a summer's afternoon. Here the vertical air currents are marked by the growth of cumulus or cumulonimbus clouds. Also, air moving from higher latitudes over warmer surfaces becomes heated from below and unstable.

(*ii*) *Cooling from above* may occur. If the upper layers of an air mass are dry, heat is radiated more quickly than from the lower moister layers. Consequently there is a change in lapse rate and a decrease in stability. Also, the tops of clouds may radiate heat during the evening while their bases are still receiving heat from a sea warmed during the day. This leads to the development of instability and thunderstorms.

(*iii*) *A conditionally unstable air mass*, that is, one with its lapse rate between the D.A.L.R. and S.A.L.R., develops instability when lifted. If it is initially lifted, say, by topography or a front, it cools at the D.A.L.R. to the condensation level, but after that, it cools at the S.A.L.R. and continues to rise under free convection. The initial lift is termed a 'trigger' action and results in complete or partial convection.

(*iv*) *The lifting of an air mass* which has a high vapour content in its lower layers but is dry aloft will also lead to instability. Initial lifting cools the lower saturated layers at the S.A.L.R. but the drier upper ones cool at the higher D.A.L.R. with a consequent change in stability.

In the case of (*iii*) and (*iv*) above, the initial lifting may be caused by any of the other processes and the latent instability of an air mass is 'triggered' off. Most occurrences of precipitation or cloud involve several of the processes, and it is not always possible to separate out the prime cause of the phenomena. Mountain ranges, by reason of their permanence, often act as 'triggers' and are associated with heavy and frequent precipitation rates, especially when athwart prevailing, maritime, air streams of latent instability.

14.4 The condensation of water vapour. Once an air mass is saturated, the next step in the precipitation process is condensation or the aggregation of the water molecules into drops of water or ice crystals. This does not always happen in a saturated air mass and super-saturation may occur with the persistent inflow of water vapour and its continued lifting. As with the case of super saturated solutions, nuclei can assist the change of state and, in air, the formation of droplets and ice. These nuclei may be dust particles, originating from the earth's surface or from outer space as in meteorites, salt crystals derived from sea spray, and combustion products. Under most conditions, sufficient nuclei are present to assure condensation when the air has reached saturation.

The drops of water and ice crystals formed in condensation are transported and supported by the vertical wind currents and are first discernable as clouds. These vertical currents vary from a gentle up-rising of 0·05 kt (0·026 m/sec) in widespread synoptic situations to vigorous updrafts exceeding 20 kt (10 m/sec) in cumulus type clouds. The stronger currents support larger drops or crystals. However, there is a wide range of drop sizes in a cloud, together with considerable variations in vertical velocities both at different heights and in different locations. A current of 0·05 kt (0·026 m/sec) supports a drop about 0·004 in. (0·1 mm) dia. and the 20 kt (10 m/sec) updraft a drop 0·04 in. (1 mm) dia. Stronger currents support drops up to 0·12 in. (3 mm) dia., but bigger water drops tend to break up. Ice crystals, by constant ascent and descent, and by experiencing strong vertical currents, can grow to much larger sizes under suitable conditions. Hail stones in excess of 3 in. (76 mm) dia. have been recorded with the largest being 7½ lb (3·4 kg) in weight. These are freak conditions however, and are not normally encountered.

14.5 Cloud particles to precipitation. With the continued inflow of moist air and vertical motion giving saturation and condensation, there is an accumulation of water droplets and ice crystals within the cloud. The cloud elements are supported by the vertical currents until they grow to a size or come to a section of the cloud when they are too heavy for the current within that part of the cloud. They now fall from the base of the cloud, and if not evaporated, reach the ground as precipitation.

The growth of the cloud droplets and ice crystals to precipitation size is the critical phase of the precipitation process, and various explanations of it have been given. The presence of ice nuclei, or crystals of similar structures and size, has been found important, especially in supercooled water clouds down to − 15° C. Also, the relative motion of drops of different sizes in the up and down drafts results in frequent collisions and the growth of droplets. The presence of larger drops appears to start coalescence and suggests that cloud seeding with appropriate sized particles assists the process of warm rain as well as that of freezing rain. It is also possible that electric charges or fields may affect the efficiency of collisions. The accretion or collision process operates at any temperature but freezing nuclei are most effective at − 15° C.

14.6 Factors involved in the precipitation intensity. The rate at which water droplets or ice crystals fall out of an air mass or, the precipitation intensity, is determined from our knowledge of the precipitation process. The relevant factors are:

(*a*) The rate at which water vapour is supplied to the storm which is producing the precipitation. This continued supply of moisture replaces that processed out of the air mass by the storm mechanism. It follows that the precipitable water of an air mass, W, and the inflow velocity into the storm, v, give the moisture supply, that is $W . v$.

(*b*) The rate at which moisture is converted from vapour to water droplets or ice particles of a sufficient size that they reach the rain gauge. This involves the rate at which the air is cooled and is proportional to the lift or the vertical velocity, w. Also involved is the rate of condensation and rate of growth of these droplets into drops of sizes which fall through the vertical currents. In turn, this depends upon the presence of suitable condensation nuclei, vertical velocities, turbulence, electrical charges, lapse rates, and other factors. It is assumed that satisfactory conditions for condensation and growth exist where saturation has been reached and that the significant factor is the vertical velocity.

(*c*) The existence of a suitable storm mechanism which, for the duration of the storm lifts the inflowing air and extracts a proportion of the available moisture. The storm possesses a distinctive flow pattern or geometry which is related to the lifting process, that is to the nature of the front, turbulence, topography, or convergence. Vertical velocities or alternatively, the net inflow in the storm, that is inflow minus outflow, are part of the flow pattern, and can be deduced when the flow pattern is known. As will be seen later this determines the proportion of water vapour which can be removed or the efficiency of the storm.

14.7 Determination of precipitation rate using vertical velocities. If a saturated air mass has a density of ρ g/cbm, vertical velocity w mps, mixing ratio r g/g, and specific humidity q g/g, and satisfactory condensation and growth conditions exist and do not vary, then the precipitation rate in the period i in./hr,

$$i = \int w_z \, \rho \, \frac{\partial q}{\partial z} \, dz. \tag{14.1}$$

Integration of this gives

$$i = \frac{w_c \rho_c \, (r_c - r_z)}{7}, \tag{14.2}$$

where subscript c represents the condensation level r_z g/g is mixing ratio at top of the lifted layer, vertical velocity is zero at top of the lifted layer, w_c mps is vertical velocity at condensation level, ρ_c g/cm^3 is air density at condensation level.

The vertical distribution of temperature and humidity is obtained from radiosondes or estimated from surface conditions. Vertical velocities however, are not usually available nor easily determined. They have been measured near the surface with cup and hot wire anemometers, and in the free air by balloons, and by the drift of light objects using radar and aircraft, but rarely to the degree of accuracy required or, in storm situations. For these reasons the formulae have few direct practical applications although indirect computations of w from other variables such as horizontal wind convergence enable estimates of precipitation intensity to be made.

As a simple example, rainfall intensity can be computed from a given vertical velocity, say, 1 mps, moisture content approximating to a pseudo-adiabat of 20°C with condensation level below 3000 ft, average air density of $1{\cdot}1 \times 10^{3}$ g/cbm, r_c of 12×10^{-3} g/g and $r_z = 0$.

Using above equation (14.2)

$$i = \frac{1 \times 1{\cdot}1 \times 10^{3} \times 12 \times 10^{-3}}{7}$$

$$= 1{\cdot}9 \text{ in./hr } (48 \text{ mm/hr}).$$

This is not an unusual rate for a convective type storm. A velocity of 10 mps would give rainfall intensities of 19 in. (480 mm) per hr; a rate which might occur over a very small area or for a period of a few minutes only. It is probable that strong currents not only form rainfall at a high rate, but also stop the drops from falling in the immediate area and distribute the precipitation formed over a wider area than where generated with consequently lower rainfall intensities.

It is possible to estimate the probable vertical wind velocities over areas by measuring the rainfall intensities and vertical humidities and using these and similar formulae. The following Table 14.1 gives estimates of the average vertical velocities and average size of water droplet particles which may be supported by these velocities in certain meteorological situations.

Table 14.1. *Types of precipitation, drop sizes and vertical velocities in meteorological situations*

Met. situation.	Av. vert. vel		Av. size of water droplets supported		Type of precipitation
	m/s	kt	mm	in.	
Widespread uplift,					
Depressions on synoptic scale	0·026	0·05	0·1	0·004	Steady light rain
Depression with layer clouds	to 0·05	to 1	0·1–1·0	0·004–0·04	Moderate rain
Convection,					
convective clouds	to 10·3	to 20	1·0–2·5	0·04–0·1	Showers
Cumulo nimbus clouds and thunderstorms	to 20·6	to 40	3·05	0·12	Heavy showers

The vertical currents and sizes of drops supported vary greatly in particular storm situations especially under unstable conditions. That is, there is a spectrum of vertical velocities and of the drops which these currents support. The figures quoted are average conditions only, and, in the storm area, many variations from these values would be found. Drops can also build up in one part of a cloud where there are strong currents and if carried to another part where there are much weaker ones, will fall out as precipitation. This is well illustrated by Fig. 14.1, the model of a hail storm, proposed by Browning and Ludlum.[5]

The rainfall equations (14.1) and (14.2) essentially comprise a convergence factor ρ times w, multiplied by a moisture factor, r or q. The two factors appear in all rainfall intensity formulae although in different guises.

14.8 The equation of continuity and precipitation rates. In view of the difficulties encountered in measuring and using the vertical velocities, the equation of moisture

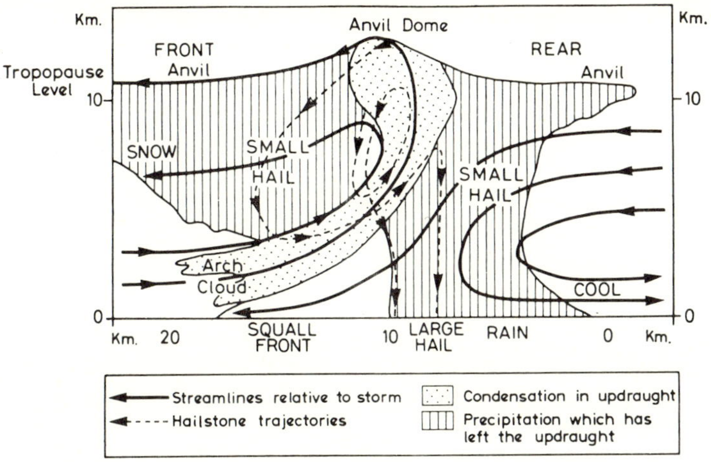

Fig. 14.1. Air flow in a typical hailstorm.

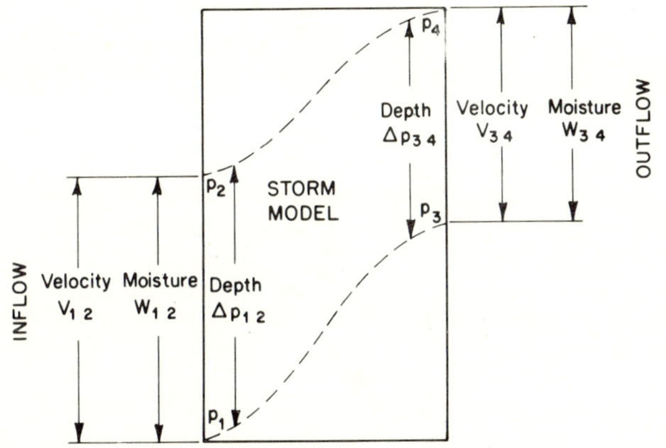

Fig. 14.2. Air flow into a storm model.

continuity is a better approach in determining rainfall intensities. The principle of continuity is applied to the flow of moisture into the space above an area A shown by Fig. 14.2. Here, in a period of time from t_1 to t_2, the moisture deposited or the rainfall intensity is equal to moisture inflow minus moisture outflow plus or minus change in moisture storage above A. The first two factors of inflow and outflow, assuming water is in the vapour phase, can be computed from horizontal velocities (v) and precipitable water (W). The changes in moisture storage above A arise from changes in the water vapour above A, measurable as the difference of W at times t_1 and t_2, plus the change in the suspended liquid water and ice crystals between times t_1 and t_2. This is discussed in paragraph 14.2 and, except when $t_2 - t_1$ is small, the change in storage above A from t_1 to t_2 is usually small in comparison with the net inflow. That is, storage is neglected in most situations encountered, for instance those when the rainfall is of more than 5 min duration. Over short periods of time, the change of storage can be the bigger part of the moisture available for deposition and makes a sizable contribution to the precipitation rate. Cloud bursts and intense falls over small areas for short durations come within this category.

The moisture can be thought of as being removed by a storm mechanism which takes into account the vertical and horizontal wind structure, the moisture charge, and the space into which the moisture is fed and released. It should be noted that the vertical and horizontal wind patterns are interrelated, but because the latter are better documented and described they are more commonly used for obtaining precipitation rates. Although the storm mechanism or model is three dimensional, it can be reduced to two dimensions with a vertical cross section representing the flow pattern.

The continuity equations of moisture and air flow are used to obtain expressions for the moisture released by a given storm model. This is shown diagramatically in Fig. 14.2 where, in unit time, moisture flows into a storm model through the inflow layer from p_1 to p_2, where the precipitable water

$$W_{12} = \int_{p_1}^{p_2} q \, dp$$

and the inflow velocity is v_{12}, evaluated over the inflow layer $p_1 - p_2$ or Δp_{12}. In unit time, moisture flows out through the layer p_3 to p_4 where precipitable water

$$W_{34} = \int_{p_3}^{p_4} q \, dp$$

and outflow velocity is v_{34}, evaluated over the outflow layer $p_3 - p_4$ or Δp_{34}. From continuity of air flow the inflow mass = outflow mass or $v_{12}\, m_{12} = v_{34}\, m_{34}$ where m represents mass but $m = \Delta p/g$ if Δp is in mb or dyn/cm^2. Therefore

$$v_{12}\, \Delta p_{12} = v_{34}\, \Delta p_{34} \quad \text{or} \quad \int_{p_1}^{p_2} v \, dp = \int_{p_3}^{p_4} v \, dp$$

or

$$v_{34} = \frac{\Delta p_{12}}{\Delta p_{34}}\, v_{12} \tag{14.3}$$

$$\text{outflow velocity} = \frac{\text{extent of inflow}}{\text{extent of outflow}} \,.\, \text{inflow velocity.}$$

From continuity of moisture flow, the moisture released from the model in unit time is inflow moisture – outflow moisture

$$\begin{aligned}
\text{released moisture} &= \int_{p_1}^{p_2} vq\,dp - \int_{p_3}^{p_4} vq\,dp \\
&= v_{12} \int_{p_1}^{p_2} q\,dp - v_{34} \int_{p_3}^{p_4} q\,dp \text{ approx.} \\
&= v_{12} \cdot W_{12} - v_{34} \,.\, W_{34} \qquad (14.4) \\
&= v_{12} \,.\, W_{12} - \frac{\Delta p_{12}}{\Delta p_{34}}\, v_{12} \,.\, W_{34} \text{ using 14.3,}
\end{aligned}$$

$$= v_{12} \cdot \left(W_{12} - \frac{\Delta p_{12}}{\Delta p_{34}} W_{34}\right) \tag{14.5}$$

$$= v_{12} \cdot W_{12} \left(1 - \frac{\Delta p_{12}}{\Delta p_{34}} \cdot \frac{W_{34}}{W_{12}}\right). \tag{14.6}$$

The expression $W_{12} - \frac{\Delta p_{12}}{\Delta p_{34}} W_{34}$ is sometimes called the effective precipitable water, W_e, for the storm model in question. It is the amount of precipitable water which can be released by the storm mechanism as distinct from the amount W_{12} which flows into it. In words formula 14.6 can be stated as the moisture released is proportional to the inflow velocity, v_{12}, the precipitable water in the inflowing air, W_{12}, and a convergence factor

$$\left(1 - \frac{\Delta p_{12}}{\Delta p_{34}} \cdot \frac{W_{34}}{W_{12}}\right)$$

which is always less than one. This latter might also be termed an efficiency factor as, for a given dew point and inflow velocity, it gives the fraction of precipitable water released. For high efficiency the negative portion of 14.6 should be small. This occurs when small amounts of moisture leave the cell as compared to the amounts that enter it. For example, inflow at sea level implies large W_{12} and, outflow at a considerable height, small W_{34}, with a high proportion of the moisture being extracted. A coastal thunderstorm is an example of such a model. Also, if the respective proportions of inflow and outflow are varied, so that $\Delta p_{12} / \Delta p_{34}$ is small, say by specifying inflow or convergence, into a limited volume but outflow or divergence over a much larger volume, then high storm efficiencies will result. Observations of storms which give high rainfall intensities such as tropical cyclones and tornadic squalls show this type of structure.

14.9 Rainfall intensity equation. In order to fully compute the amount of moisture released in a storm model and convert it to a precipitation depth, it is necessary to know the additional factors of the geometry of the storm model and the basin over which the precipitation falls. For instance, the amount of moisture supplied depends upon the width of inflow, the proportion released depends on the storm model, and the depth of precipitation depends upon the area over which the released moisture is distributed.

If K represents a geometric factor, the rate of precipitation, i, can be expressed as

$$i = K v_{12} \left(W_{12} - \frac{\Delta p_{12}}{\Delta p_{34}} W_{34}\right) \tag{14.7}$$

or

$$i = K v_{12} \; . \; W_e \tag{14.8}$$

Taking account of variations with time, the total precipitation depth from time t_1 to t_2 is P and

$$P = \int_{t_1}^{t_2} K.v.W_e \; dt \tag{14.9}$$

more generally

$$P = \int fv \, dt \tag{14.10}$$

where f is a combined moisture factor which is dependent on geometry, convergence, and precipitable water inflow. These last four equations indicate that the depth of precipitation and the rainfall intensity, depend upon a geometric factor, K, inflow velocity v, precipitable water of inflow, W_{12} , and a storm convergence or efficiency factor,

$$\left(1 - \frac{\Delta p_{12}}{\Delta p_{34}} \cdot \frac{W_{34}}{W_{12}}\right)$$

If the inflow velocity is considered as part of the storm model and incorporated into the convergence factor, the rainfall intensity can be expressed in terms of a moisture factor and one overall convergence factor. The convergence factors can be determined for actual storms, without defining storm geometry or inflow velocities, from the relationship between rainfall intensity and moisture content. [6] Thus, it is possible to maximise actual intense storms which are assumed to have reached their maximum efficiency, or highest convergence factor by using the ratio of the highest recorded moisture content to the actual moisture content of the storm. For the efficient or intense storm maximisation of the moisture factor gives an estimate of the extreme rate of rainfall.

REFERENCES

[1] MASON B.J., (1957). *Physics of Cloud*, Clarenden Press, Oxford.

[2] REITAN C.H., (1960). Distribution of precipitable water vapour over the continental United States. *Bull. Am. Met. Soc.* **41**, 78–87.

[3] WIESNER C.J., (1964) . Ed., *Water Resources Uses and Management*, p. 290 , Melbourne University Press.

[4] KNOX J.B., (1960) . Procedures for estimating maximum possible precipitation. *Bull No.88 Dept. of Water Resources*. California.

[5] BROWNING K.A. and LUDLUM E.H., (1962) . Air flow in convective storms. *Q. J. R. Met. Soc.* **88**, 117–135.

[6] FLETCHER R.D., (1948) . Computation of thunderstorm rainfall. *Trans. Am. geophys. Uni.* **29**, 41–50.

CHAPTER FIFTEEN

Storm models

15.1 Use of storm models. The study of the precipitation process and the factors involved in precipitation rates give a basis for setting up flow patterns, storm models, or storm mechanisms. From such patterns it is possible to establish formulae connecting the variables involved, and to estimate the precipitation rates under specified conditions of vertical lift, moisture content, and wind flow. This technique enables theoretical computations to be made of storm rainfall. It is a logical method of extrapolating precipitation to its probable extreme, or maximum value, by using the likely maximum values of the component elements in the precipitation formulae.

Storm models or flow patterns may be set up in three dimensions and can allow for the presence of moisture barriers, variations in depths and extent of inflow and outflow, directions of inflow and outflow with allowance for cyclonic curvature, the simplified topography of an area, and any storm or basin characteristics. In most cases a vertical cross section can reasonably represent the storm's flow pattern into a project basin. This reduction to two dimensions simplifies the calculations and reveals the relevant factors in storm precipitation.

In natural storms, especially the severe ones which produce intense rainfall, it is difficult to obtain observations by which the theoretical pattern can be verified. Most of the early upper wind observations were made visually and were not available during storms. Winds observed by radio methods give a better description of the flow but observations are few and at widely scattered stations. The path traced by a balloon is a minute section of the flow pattern over a limited time period, whereas knowledge of the entire storm structure and of the sequence of events which occur within it is required. Radar promises to give more complete pictures of the circulations as witnessed by the description of a hailstorm by Browning and Ludlum [1] and shown in Fig. 14.1. An additional difficulty is that vertical wind velocities which are decidedly relevant in storm studies, cannot be accurately measured over the wide areas covered by storms.

Simple storm models permit the isolation of significant factors during heavy rates of precipitation. These models may incorporate coefficients for specific catchments to take account of the basin's geometry. In order to decide on the particular storm model which should be applied to a valley, the intense storms which have occurred within the homogeneous meteorological zone of which the valley forms part, must be analysed to isolate the meteorological parameters concerned with precipitation rates. A storm model based on these analyses and devised for a particular catchment, can be set up and tested for actual situations. These tests indicate whether it represents conditions or needs modification. A satisfactory model permits estimations of extreme precipitation rates to be made by inserting the highest significant values of the parameters which can coexist within the rainfall equation of the storm model.

The calculation of precipitation rates from storm models depends upon the principle of continuity of air flow and of moisture flow, but many simplifying assumptions are made because of the lack of data. For example, the storage of liquid and solid water within an air mass cannot be successfully measured before and after rain and so is usually neglected. Also storm models are more successful in large scale situations rather than in small intense storms mainly because errors in the moisture flow in a large storm are a smaller proportion of the total inflow and precipitation. Also, storage is less important. Usually orographic models are the most satisfactory as the vertical lift is related to the permanent physical features and therefore the flow pattern and the wind velocities are well defined. The two simplest models are those of upglide and convergent inflow. The upglide model describes the processes which are encountered in topographic and frontal rains while the inflow model describes the flow patterns commonly found in storms involving convergence and turbulence.

15.2 The upglide model. A simple way of lifting a stream of air allows it to impinge on an inclined plane facing upwards. In nature, the inclined plane is a mountain barrier, or the frontal surface between two air masses where the warmer air is lifted by, or rides over, the lower dome of cold air. The process is described by an air mass, containing precipitable water W_{12}, entering a rectangular valley X wide and Y long with an inflow velocity v_{12} and leaving the end of the valley, after a uniform rise to a height z, with outflow velocity v_{34} and precipitable water W_{34}. This is shown in Fig. 15.1.

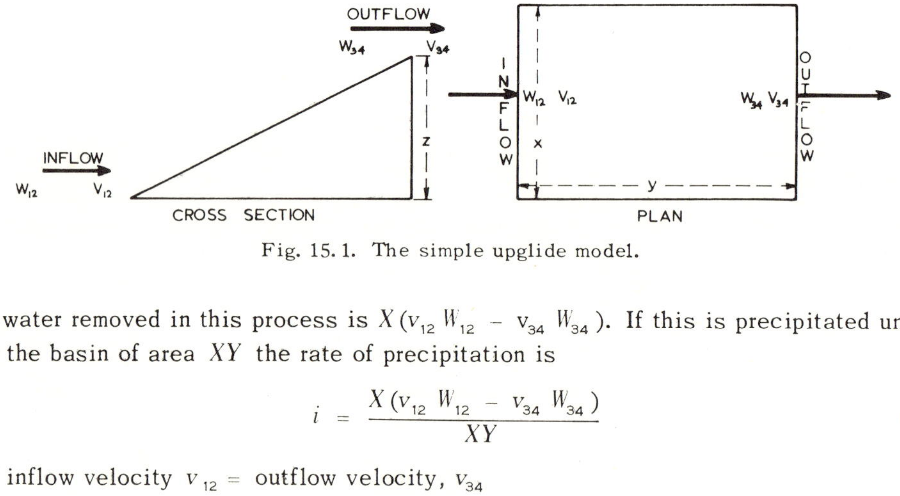

Fig. 15.1. The simple upglide model.

The water removed in this process is $X(v_{12}W_{12} - v_{34}W_{34})$. If this is precipitated uniformly over the basin of area XY the rate of precipitation is

$$i = \frac{X(v_{12}W_{12} - v_{34}W_{34})}{XY} \tag{15.1}$$

or if inflow velocity v_{12} = outflow velocity, v_{34}

$$i = \frac{v_{12}(W_{12} - W_{34})}{Y}. \tag{15.2}$$

The precipitation rate depends upon a geometric factor $1/Y$, also termed the basin constant, the inflow velocity v_{12} and a convergent factor $W_{12} - W_{34}$ which depends upon the precipitable water entering the valley W_{12} and the lift in the basin

$$\left(1 - \frac{W_{34}}{W_{12}}\right), \text{ a function of } z.$$

If the vertical extent of inflow and outflow are considered in Fig. 15.2 as occurring between the levels p_1 to p_2 and p_3 to p_4 respectively where pressure represents the height, then the precipitation equation becomes,

$$i = \frac{1}{Y}\left\{\int_{p_1}^{p_2} vq\,dp - \int_{p_3}^{p_4} vq\,dp\right\} \tag{15.3}$$

or

$$i = \frac{1}{Y}\,(v_{12}\,W_{12} - v_{34}\,W_{34}) \tag{15.4}$$

where v_{12} , v_{34} , and W_{12} , W_{34} represent wind velocities and precipitable water throughout inflow and outflow.

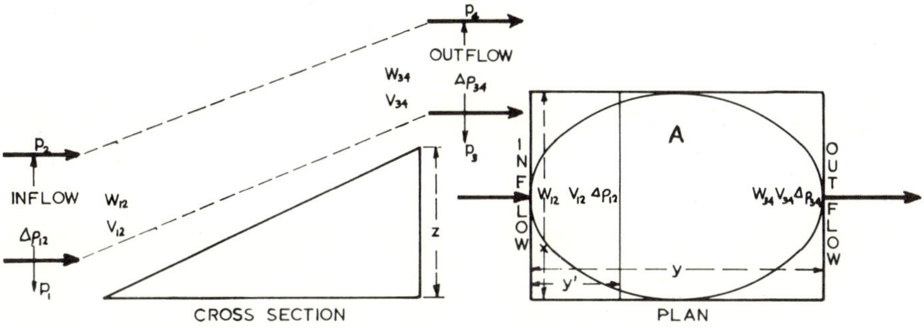

Fig. 15.2. Upglide model with vertical extent.

It follows from continuity that $v_{12}\,\Delta p_{12} = v_{34}\,\Delta p_{34}$ (15.5)

$$i = \frac{1}{Y}\,v_{12}\left(W_{12} - \frac{\Delta p_{12}}{\Delta p_{34}}\,W_{34}\right) \tag{15.6}$$

$$i = \frac{1}{Y}\,v_{12}\cdot W_{12}\left(1 - \frac{\Delta p_{12}}{\Delta p_{34}}\cdot\frac{W_{34}}{W_{12}}\right). \tag{15.7}$$

Thus, rainfall intensity depends on a geometric or basin factor $1/Y$, inflow velocity, v_{12} , a moisture factor, W_{12} , and a convergence factor. In the convergence factor there is the geometry of the storm model in the terms

$$\frac{\Delta p_{12}}{\Delta p_{34}} \quad \text{and} \quad \frac{W_{34}}{W_{12}} .$$

If, instead of the water being deposited over the rectangle XY, it is deposited over an area A, the basin constant becomes X/A. This area A may be an ellipse within the rectangle XY and, with basin constant X/A or, it may be a smaller rectangle, X by Y^1, with basin constant $1/Y^1$.

It might be noted that, although it is not possible for all the precipitable water flowing into a storm model to be deposited out, the expression $1/Y\ v_{12}\ W_{12}$ represents a precipitation rate which could not be exceeded.

A simple example postulates that saturated air of dew point 70°F flows at right angles into a rectangular valley 30 miles wide and 50 miles deep, at a speed of 20 miles/hr. This air mass extends from 1000 to 300 mb and leaves the valley at 3000 ft (approx. 900 mb) with outflow extending to 300 mb.

Assuming a saturated pseudo-adiabatic lapse rate and uniform rain over the basin, the rainfall intensity is

$$i = \frac{30 \text{ miles} \times 20 \text{ miles/hr}}{30 \text{ miles} \times 50 \text{ miles/hr}} \left(2{\cdot}26 - \frac{1000 - 300}{900 - 300}\, 1{\cdot}67\right)$$

where precipitable waters are obtained from Fig. 13.2 or Tables.

$$i = \frac{20}{50}\,(0{\cdot}31)$$

$$= 0{\cdot}12 \text{ in. per hr.}$$

If the rain were deposited in the first 10 miles of the valley, then the basin constant would be

$$\frac{30}{30 \times 10} = \frac{1}{10}$$

and $$i = \frac{20}{10}\,(0{\cdot}31)$$

$$= 0{\cdot}62 \text{ in. per hr.}$$

Again, if the basin has an elliptical area of 1200 square miles, and rain is uniform over it, then the basin constant is

$$\frac{30}{1200} = \frac{1}{40}$$

and $$i = \frac{20}{40}\,(0{\cdot}31)$$

$$= 0{\cdot}15 \text{ in. per hr.}$$

If all the precipitable water could be released and deposited over the basin, the rate would be seven times the above rates.

15.3 The circumscribing rectangle. Basins or catchments are not naturally rectangular. However, a rectangle may be circumscribed about a basin so that one side, X, is normal to the other, Y, which can be made parallel to the inflowing air. Basin constants are then X/A or $1/Y$. Thus, basin constants depend upon the wind direction and a series of circumscribing rectangles and corresponding basin constants may be determined for all significant wind directions into the basin. The amount of inflowing precipitable water is also dependent on wind direction. In a particular situation, wind velocities, dew points, orographical barriers, and depths of air masses vary according to the direction of the wind.

Detailed geometrical analyses, based on past storms, show the directions which are significant and which give the highest precipitation rates in the important storms of the catchment under study.

15.4 Other processes within an air mass. The simple upglide mechanism assumes that the air mass is lifted from p_1 to p_3 without other processes such as convergence or convection occurring within it to cause further lift. The air mass is assumed stable and the precipitation rate results from the lifting of this stable air mass. In nature, such processes are uncommon and give light rates of precipitation. Usually, upglide is accompanied by convergence and convection. For instance, when orographic uplift occurs, the lifted air mass often becomes unstable with further ascent. Precipitation due to convective uplift is added to that produced by upglide.

15.5 The convergence model with radial inflow. If air is forced to converge within an area, vertical motion must ensue. The vertical velocity is dependent upon the rate of convergence which is measured by the decrease of horizontal area with time.

A simple case is one of radial inflow when air of precipitable water, W_{12} , moves radially with velocity v_{12} into a circular column of radius r as shown in Fig. 15.3. The total inflow of moisture over the perimeter of the circle, $2\pi r$, is $2\pi r\ v_{12}\ W_{12}$. If all this moisture were deposited over the base of the cylinder, πr^2, the rainfall intensity

$$i = \frac{2\pi r\ v_{12}\ W_{12}}{\pi r^2} = \frac{2v_{12}\ W_{12}}{r} \quad . \tag{15.8}$$

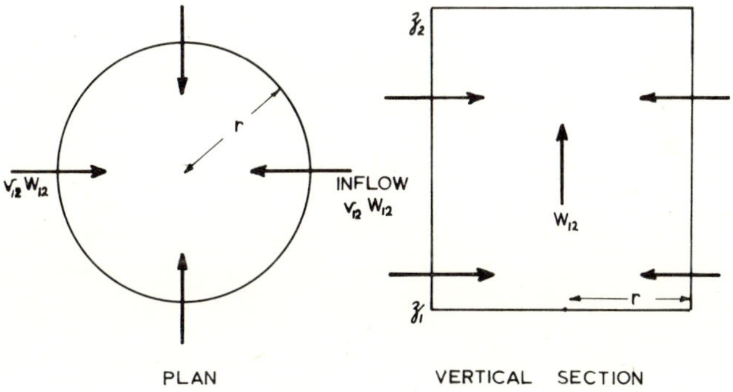

Fig. 15.3. Simple radial inflow without outflow.

Again there are a basin factor, $2/r$, inflow velocity, moisture factor, and an efficiency or convergence factor. The last is assumed at the impossible value of 1 in this simple example. There is no known atmospheric pressure situation which allows such a condition to persist although rapidly deepening cyclones, waves, and isolated storms may approach similar conditions for short periods over small areas. The process of convergence alone tends to destroy the situation producing it and to persist is necessarily accompanied by divergence or outflow. If there is no outflow, pressures rise rapidly and pressure gradients are reversed. A similar effect occurs in a thunderstorm in its dying phases. Inflow into a convergent model is sustained provided there is compensating outflow.

The vertical velocity, w_{12}, in the simple radial inflow model of height $z_2 - z_1$, can be shown to be

$$w_{12} = \frac{2(z_2 - z_1)}{r} v_{12} . \tag{15.9}$$

In verification it has been shown that if a 10 m/sec air mass moves into a 13 km radius cell then the vertical velocity at 1 km approaches 1·6 m/sec.

15.6 Radial inflow with compensating outflow. A convergence cell with inflow and outflow is shown in Fig. 15.4. If inflow of velocity v_{12} and precipitable water W_{12} occurs from p_1 to p_2 or z_1 to z_2, and outflow of v_{34} and W_{34} from p_3 to p_4 or z_3 to z_4, then the released precipitable water, or effective precipitable water W_e for the storm model is, from continuity considerations,

$$w_e = w_{12} - \frac{\Delta p_{12}}{\Delta p_{34}} W_{34} .$$

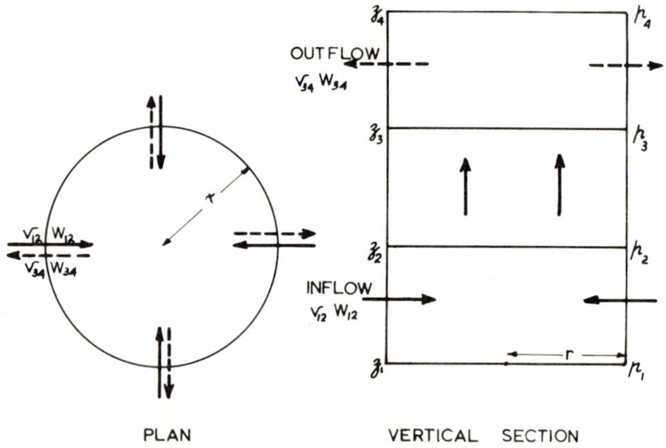

Fig. 15.4. Radial inflow with compensating outflow.

The rainfall intensity in this cell of radius r is

$$i = \frac{2v_{12}}{r}\left(W_{12} - \frac{\Delta p_{12}}{\Delta p_{34}} \cdot W_{34}\right) \tag{15.11}$$

$$i = \frac{2v_{12}}{r} W_e . \tag{15.12}$$

Factors of geometry of basin, inflow velocity, moisture, and storm model efficiency or convergence, are apparent. A simple example illustrates the significance of outflow in these formulae.

If saturated air of dew point 70°F flows into a cell 10 miles in radius at a velocity of

10 miles/hr between 1000 and 800 mb, and flows out between 600 and 400 mb, from continuity this will be at the same velocity and

rainfall intensity $$i = \frac{2}{r} v_{12} \left(W_{12} - \frac{\Delta p_{12}}{\Delta p_{34}} W_{34} \right)$$

$$= \frac{2}{10} \times 10 \left(1 \cdot 10 - \frac{200}{200} \times 0 \cdot 36\right)$$

$$= 1 \cdot 48 \text{ in./hr.}$$

If there is infinite divergence

$$i = \frac{2}{r} v_{12} W_{12}$$

$$= \frac{2}{10} \times 10 \times 1 \times 10$$

$$= 2 \cdot 20 \text{ in./hr.}$$

If outflow occurs between 500 and 400 mb, outflow velocity is doubled but outflow moisture reduced, and

$$i = \frac{2}{10} \times 10 \left(1 \cdot 10 - \frac{200}{100} \times 0 \cdot 14\right)$$

$$= 1 \cdot 6 \text{ in./hr.}$$

If outflow occurs between 800 and 400 mb, outflow velocity is halved but outflow moisture increased, and

$$i = \frac{2}{10} \times 10 \left(1 \cdot 10 - \frac{200}{400} \times 1 \cdot 11\right)$$

$$= 1 \cdot 10 \text{ in./hr.}$$

15.7 Types of convergence and convergence factors. Convergence may be considered as the horizontal shrinking of an air mass or column of air. This means that convergence may be expressed in terms of the proportional shrinking of area in unit time. For example, convergence of 4×10^{-5} per sec means that the horizontal cross section of an air column reduces in area by a factor of 0·00004 every second.

(*a*) *Radial convergence* is shown in Fig. 15.5a, as occurring when the area *ABCD* shrinks to *A′B′C′D′*.

$$\text{Rate of convergence} = \frac{ABCD - A'B'C'D'}{ABCD} = \frac{2rv - v^2}{r^2}$$

(*b*) *Down wind convergence* occurs with decreasing wind velocity down wind and is shown in Fig. 15.5b with the area shrinking from *ABCD* to *A′B′C′D′*. When the velocity at *AB* (v_{ab}) is greater than at *CD* (v_{cd}) the rate of convergence is

$$\frac{v_{ab} - v_{cd}}{BC}$$

or dependent upon the difference between down wind velocities.

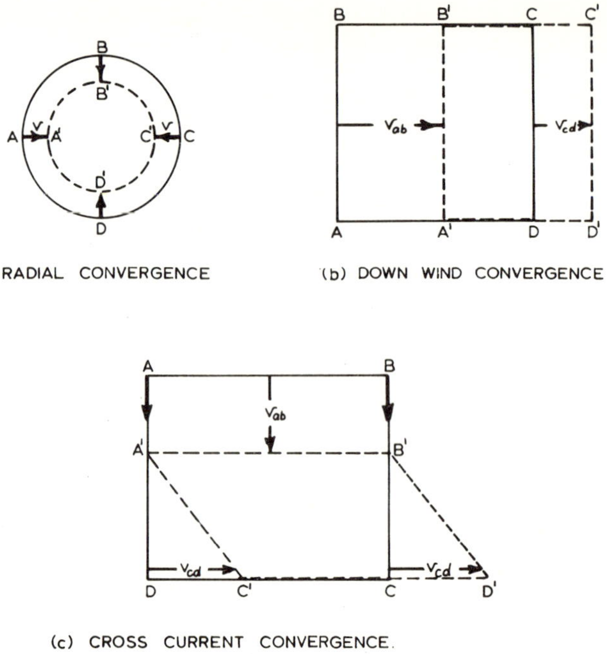

Fig. 15.5. Types of convergence.

(*c*) *Cross current convergence* occurs when one air mass moves into another at an angle as shown in Fig. 15.5c. The air mass at AB moves with velocity v_{ab} to $A'B'$ while the air at CD moves to $C'D'$ with velocity v_{cd}. The area $ABCD$ is reduced to $A'B'C'D'$ and rate of convergence is dependent on concurrent wind velocities. Necessary adjustments can be made when the wind is not at right angles.

15.8 The geometry of the convergent cell. If rates of precipitation are to be maintained in convergent cells, there must be outflow as well as inflow. In fact, the effectiveness or efficiency of a storm model in removing water from the air mass which passes through it depends upon the geometry of the cell and particularly its vertical extent and the proportions of inflow and outflow. Moisture inflow comprising inflow velocities and precipitable water are two other factors.

The convergent cell in Fig. 15.6 shows inflow between pressure levels p_1 and p_2 and outflow between pressure levels p_3 and p_4, and assumes a saturated air mass with specific humidity q varying as shown. The precipitation rate depends upon the moisture released or,

$$v_{12} \int_{p_1}^{p_2} q\,dp - v_{34} \int_{p_3}^{p_4} q\,dp$$

or

$$v_{12} \left(\int_{p_1}^{p_2} q\,dp - \frac{\Delta p_{12}}{\Delta p_{34}} \int_{p_3}^{p_4} q\,dp \right)$$

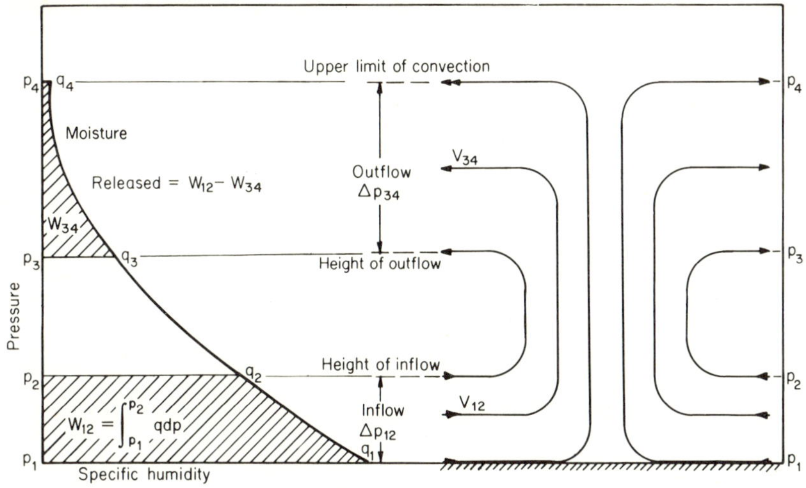

Fig. 15.6. The convergent cell.

as
$$\Delta p_{12}\ v_{12} = \Delta p_{34}\ v_{34}.$$

That is,
$$v_{12}\left(W_{12} - \frac{\Delta p_{12}}{\Delta p_{34}} \times W_{34}\right).$$

The following geometrical factors are therefore significant.

(*a*) *The upper limit of convection* (p_4). The vertical extent of the cell p_4 determines the amount of water available and the total space available for inflow, outflow and nil divergence. In nature, the height of convection is determined from observations of the tops of cumulonimbus clouds within the various latitudes and seasons. The extent of convection is related to surface temperature and is highest in the tropics and during summer. It decreases towards the poles and during winter. Observed variations are from 28,000 ft to 53,000 ft, corresponding approximately to pressure levels of 300 mb and 100 mb.

The tropopause, which also varies in height with latitude and season, marks the limit of the mixing of air from the surface layers, gives additional information on the upper limit of convection, and confirms the above figures.

The mean temperature of the ocean varies in a known way with latitude and is proportional to the surface dew points of the overlying air. Thus the upper limit of convection can be related to the surface dew point. Accepting that a reasonable range of surface dew points is from 50°F in high latitudes to 78°F in the tropics at 1000 mb, a dew point of 50°F corresponds to a height of convection of 300 mb and, a dew point of 78°F to a height of 100 mb. Again, assuming that the cell height is proportional to the vapour pressure at the surface, the upper limit of convection may be plotted against a horizontal scale of surface (1000 mb) dew point. This is shown in Fig. 15.7.

(*b*) *The depth and height of the inflow layer,* Δp_{12} *or* $p_1 - p_2$ *and* p_2. The inflow layer extends from p_1 to p_2 and forms the bottom part of the cell. In the surface layers, the higher amounts of water vapour are normally available, with the moisture content in a saturated air mass usually varying as shown in Fig. 15.6. Raising the base of the cell, p_1, reduces the available water for inflow as specific humidity drops off with height. This depletion is discussed further under inflow barriers.

Besides being a function of the storm model, the depth and height of inflow varies with the upper limit of convection and the surface dew point. This variation is shown in Fig. 15.7.

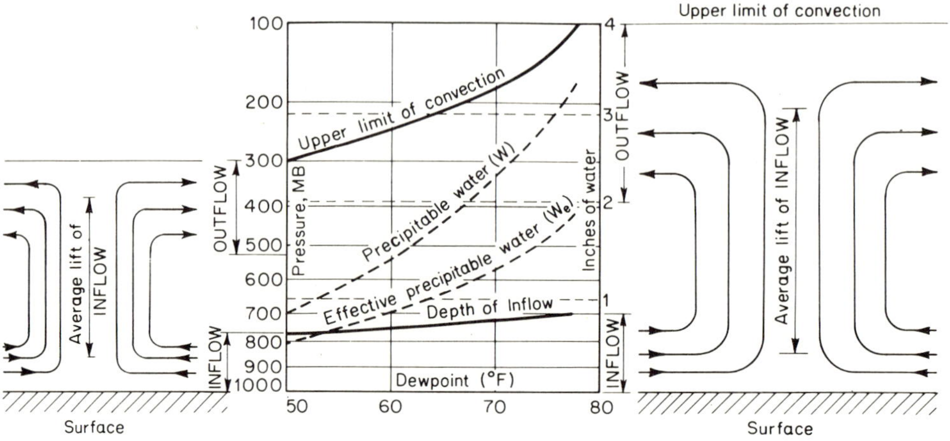

Fig. 15.7. Convergent cell in relation to surface dew point and its geometry.

The depth of inflow is related to the depth of outflow by the continuity equation

$$v_{12} \times \Delta p_{12} = v_{34} \times \Delta p_{34}.$$

If three of these are known, then the fourth is also determined. Normally inflow velocity, and the proportions of inflow and outflow, are known or assumed.

The shape of the cumulonimbus cloud suggests a vertical division of the convective cell into three layers of equal pressure differences, with inflow occurring in the lower third, vertical motion in the middle section and outflow out of the top third. This is the generally accepted cell with inflow velocity equalling outflow velocity.

(*c*) *The depth and height of the outflow layer,* Δp_{34} *or* $p_3 - p_4$ *and* p_4. The outflow layer extends from p_3 to p_4 and is the top portion of the convergent cell. Temperatures and moisture contents are lowest in this section. The higher the outflow layer the more water can be released by the cell.

As with inflow, the height and depth of the outflow layer depends upon the storm model itself, the height of convection and, the surface dew point. For the equally proportioned cell, the outflow layer is shown in Fig. 15.6 and 15.7.

(*d*) *The zone of nil convergence or divergence, (from* p_2 *to* p_3*).* Between the inflow and outflow layers is a zone where there is neither horizontal convergence nor divergence. This is the zone where vertical motion or lift is taking place and moisture may be stored or held in its liquid or solid form. Although this is often made the same size as the inflow and outflow sections, as suggested by photographs, there is no reason why this should be so and various other proportions may be assumed or, the centre section entirely eliminated. Outflow, however, must always balance inflow. The average lift of the inflow layer is

$$\frac{p_1 + p_2}{2} - \frac{p_3 + p_4}{2}.$$

(*e*) *The relative proportions of inflow and outflow,* $\Delta p_{12}/\Delta p_{34}$. The depths of inflow and outflow will be related to inflow and outflow velocities by means of the equation of continuity. A high velocity inflow may occur through a limited depth and be associated with low outflow velocities throughout greater depths aloft.

The water removed is proportional to

$$v_{12}\left(W_{12} - W_{34}\frac{\Delta p_{12}}{\Delta p_{34}}\right)$$

and although an increase in Δp_{34} at the expense of Δp_{12} gives a smaller proportion of W_{34} flowing out, this involves lowering the level p_3 and so increasing W_{34}. Similarly, lowering p_2 gives a decrease in W_{12}. Tests of cells of different proportions show that more water is usually released when the inflow depth exceeds the outflow depths. Again, in small storms, a large inflow depth may give high vertical velocities which do not permit the condensed water to fall in their vicinity. It appears that the three unit cell is a reasonable compromise and is usually assumed as applying in nature.

15.9 The surface dew point and the geometry of the convergent model. In the three unit cell of 1/3 inflow, 1/3 vertical rise and 1/3 outflow, as shown in Fig. 15.7, the geometry of the cell varies with the surface dew point. The relationship is shown between the 1000 mb surface dew point and the following factors:

(*a*) The upper limit of convection and the precipitable water in the layer below.

(*b*) The height and depth of the inflow layer and the precipitable water flowing into the cell.

(*c*) The height and depth of the outflow layer and the precipitable.water flowing out of the cell.

(*d*) The average lift of the inflow layer.

(*e*) The amount of water released by the cell, defined at the dew point. This is the effective precipitable water.

15.10 Effective precipitable water W_e. The amount of precipitable water, W_e, released by a convergent cell with inflow p_1 to p_2 and outflow from p_3 to p_4 is

$$W_e = W_{12} - \frac{\Delta p_{12}}{\Delta p_{34}} \cdot W_{34}.$$

This expression is generally termed the effective precipitable water, W_e, and depends not only on surface dew points, but also on the geometry of the cell. A better term might be the released precipitable water. As the amount released depends upon the storm model, as well as the moisture content of the air, it can be used as a measure of storm efficiency. The convective cell is an efficient mechanism, and, assuming equal proportions of inflow, updraft and outflow, a value $W_e = W_{12} - W_{34}$ is obtained which is often specifically termed the effective precipitable water. This was used extensively in early hydrometeorological studies. The variation of W_e with surface dew point is shown in Fig. 15.7. This restricted definition of W_e should be distinguished from the general term of effective precipitable water used for all models and which might better be called released precipitable water.

In Table 15.1 for an air mass of dew points 50, 60, 70 and 78°, saturated at the pseudo-adiabatic lapse rate, are given the precipitable water and the effective precipitable water using the equally proportioned cell, and the recommended heights of convection, inflow and outflow.

Table 15. 1. *Factors in the equally proportioned convergent cell*

Dew pt °F	Upper limit of convection p_4, mb	Inflow W_{12} (in.)	Outflow W_{34} (in.)	Released $W_e = W_{12} - W_{34}$ (in.)	Total W (in.)	Storm efficiency $\frac{W_e}{W} \times 100$ (%)	$\frac{W_e}{W_e}$ at 78°	$\frac{W}{W}$ at 78°
50	300	0·55	0·04	0·51	0·84	61	0·268	0·251
60	240	0·89	0·07	0·82	1·38	59	0·432	0·408
70	150	1·39	0·10	1·29	2·27	59	0·679	0·672
78	100	2·07	0·17	1·90	3·35	57	1·000	1·000

It is also possible to determine a storm efficiency factor or, the proportion of water removed by the cell, namely W_e/W. In the same table, the ratio of the effective precipitable water at the given dew point to the effective precipitable water at a dew point of 78° are given and compared with a similar ratio for precipitable water. These ratios are sufficiently alike to warrant use of the ratio of precipitable water instead of the ratio of effective precipitable water wherever the latter occurs. Precipitable water is more easily computed than effective precipitable water. The importance of this is seen in the maximisation and transposition of storms, which is considered later. Storms are adjusted for moisture content by using either values of effective precipitable water or values of precipitable water.

15.11 Convergent cells of varied proportions. On visual evidence of the shape of cumuliform clouds, it is assumed that the convective or convergent cell is comprised of three layers of equal pressure differences. Other proportions of inflow and outflow occur and such cells possess higher or lower efficiencies in the removal of water from air masses. To examine this aspect of the convergent cell, theoretical computations are made with cells of different inflow and outflow by considering the total depth to comprise eighteen equal pressure intervals and ascribing a number of these units to inflow, p_{12}, a number to the central section, p_{23}, and the remainder to outflow, p_{34}. Thus the cell discussed above has proportions, 6 : 6 : 6, other proportions such as 8 : 2 : 8, 8 : 4 : 6, and 12 : 0 : 6 may be taken and, by considering a saturated air mass with 1000 mb dew points of 50, 60, 70 and 78°F, the released precipitable water

$$W_{12} - \frac{\Delta p_{12}}{\Delta p_{34}} W_{34}$$

is computed for cells rising to the limit of convection. The results of the calculations are shown in Table 15.2.

Table 15.2. *Water released from convergent cells of varied proportions*

Inflow Δp_{12}	Centre Δp_{23}	Outflow Δp_{34}	1000 mb Dew pt °F	Released water (in.)	Total water (in.)	Released / Released at 78°F	Total / Total at 78°F
Equal sections			50	0·51	0·84	0·268	0·251
6	6	6	60	0·82	1·38	0·432	·408
			70	1·29	2·27	0·679	·672
			78	1·90	3·35	1·000	1·000
Inflow equals outflow			50	0·57	As above	0·266	As above
8	2	8	60	0·91		0·425	
			70	1·45		0·676	
			78	2·14		1·000	
Inflow greater than outflow			50	0·61	As above	0·262	As above
8	4	6	60	0·98		0·421	
			70	1·57		0·674	
			78	2·33		1·000	
No central section and inflow greater than outflow			50	0·71	As above	0·251	As above
12	0	6	60	1·16		0·410	
			70	1·89		0·676	
			78	2·83		1·000	

It will be noticed that more water is released by the later models indicating that cells with these proportions are more efficient in extracting water from the saturated air. However, the ratio of the released precipitable water for each model and dew point, to the amount released by the same model at a dew point of 78°F is reasonably constant. In fact it is approximately equal to the ratio of the precipitable waters at the two dew points. This latter ratio is thus independent of the proportions of the cells and depends only upon surface dew point and the related height of convection. This means that, irrespective of the actual proportions of a storm cell, it may be adjusted for moisture content by using the ratio of the precipitable water at the respective dew points.

If actual storms are to be adjusted to allow for their occurrence in different sites or conditions, (as in the processes of transposition and maximisation), it is not necessary to specify the exact model which gives the rainfall, but only the respective dew points or precipitable waters of the air masses involved. Adjustment for greater, or less, moisture inflow may be done by using the ratio of precipitable waters. It is implied that the storm model is unchanged by transposition or maximisation.

In the adjustment of actual storms it is also assumed that the major storms which have occurred include the efficient ones with an optimum combination of inflow wind and geometrical proportions, and that these factors do not need to be adjusted further. The moisture factor only is adjusted. This is discussed later under probable maximum precipitation.

Ratios of precipitable waters are thus adjustment factors and are shown plotted against 1000 mb dew points in Fig. 15.8. This figure may be used to give adjustment

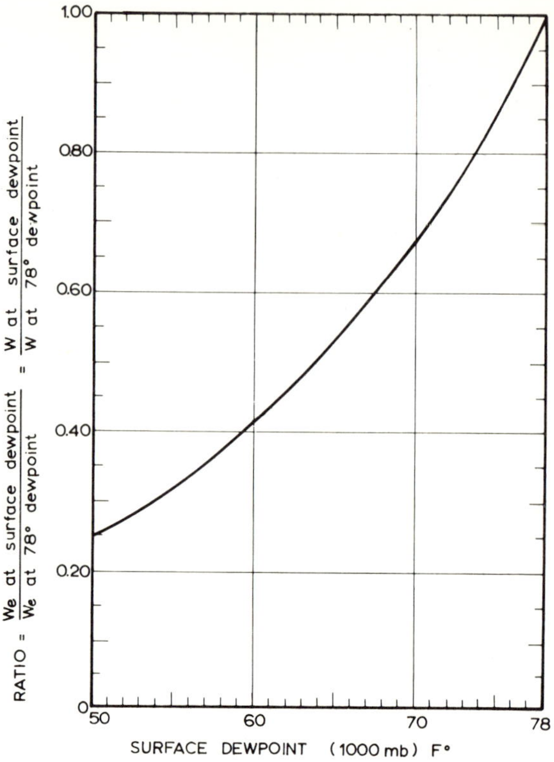

Fig. 15.8. Ratios of precipitable water at given surface dew point to precipitable water at 78°F.

factors by comparing appropriate ratios. For example, if a storm of 60°F dew point is to be adjusted to 70°F dew point, the appropriate adjustment factor is

$$\frac{\text{ratio at } 70^\circ\text{F}}{\text{ratio at } 60^\circ\text{F}} = \frac{0{\cdot}67}{0{\cdot}41} = 1{\cdot}63.$$

Alternatively Fig. 13.2 is used to determine actual precipitable waters at each dew point and then compared

$$\text{adjustment factor} = \frac{W \text{ at } 70^\circ\text{F}}{W \text{ at } 60^\circ\text{F}} = \frac{2{\cdot}27}{1{\cdot}38} = 1{\cdot}64.$$

It is of interest to note that a comparison with the effective precipitable waters gives

$$\text{adjustment factor} = \frac{W_e \text{ at } 70^\circ\text{F}}{W_e \text{ at } 60^\circ\text{F}} = \frac{1{\cdot}29}{0{\cdot}82} = 1{\cdot}57.$$

15.12 Adjustment of convergent cells for height of base. The convergent cells have been considered as occurring with their bases at 1000 mb or sea level. As these cells occur at other elevations the effect of raising the cell base is relevant.

From work already considered, it is apparent that when a storm model is lifted,

(*a*) The precipitable water is depleted with height (depletion).

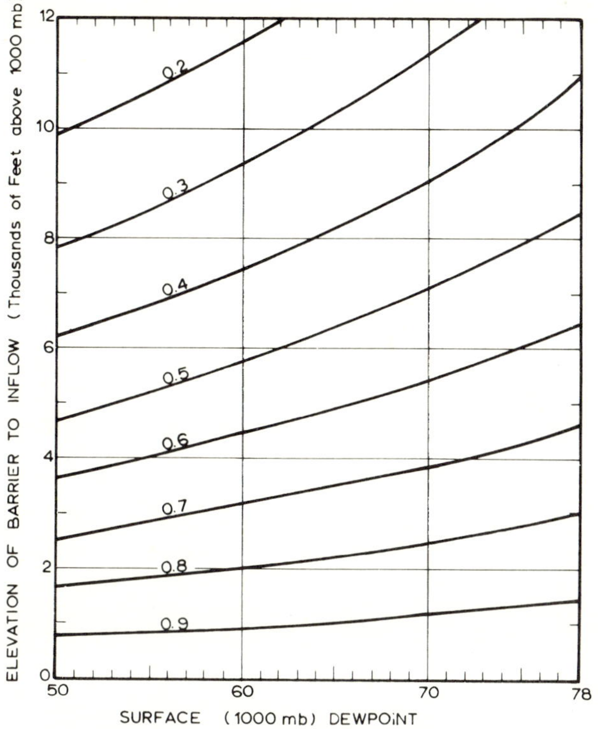

Fig. 15.9. Proportional depletion of air mass by raising base of cell.

(*b*) Lifting may intensify precipitation (intensification). Lapse rates may be changed and produce a new storm mechanism or modify the existing one.

(*c*) Precipitation induced in a cell may be carried to fall in an adjacent area (spillover).

The relative importance of (*a*), (*b*) and (*c*) depends upon the particular weather situation, or more precisely, on the storm mechanism, air mass, and the topography. In some cases the depletion is balanced by intensification while, in others, each factor can be assessed and allowed for as follows,

(*a*) Estimate the amount of depletion of W with height from Fig. 13.2.

(*b*) Assess the orographic intensification effect by comparing stations where there is no orographic lift with stations where it is present, to obtain ratios of orographic to convergence rain.

(*c*) Allow for spillover by shifting the position of the moisture barrier or by the calculation of the trajectories of raindrops or snow.

The storm and location is examined to determine whether the depletion effect cancels the intensification effect in whole or in part. After this examination, either no adjustment, or a proportional adjustment, is made for elevation. This storm analysis also indicates whether certain storms only should be transposed to regions of similar topography.

A full analysis of the storms within a basin involving allowances for depletion, intensification, and spillover should be done when data and manpower is adequate, important works are planned, or critical decisions are to be made. Although intensification and spillover depend on particular circumstances, the depletion effect always occurs and can be estimated and applied to most situations. It will be further considered.

The effect of elevating the base of convergent cells is computed by reducing the pressure at the base, while keeping the same pressure at the top of cell and fitting the proportions of the cells into the new pressure range. The height of the cell is decreased by the amount the cell base is raised and the layers within the cell are reduced in the same proportion. For example, if the base of a 1000 to 100 mb cell at 78°F dew point, is raised to 900 mb, each of the layers is reduced by 100/900. Proportions of the 6 : 6 : 6 cell originally 1000, 700, 400, 100 mb become 900, 633, 367, 100 mb. Proceeding in this way, the released precipitable water, W_e , for 6 : 6 : 6 cells with 1000 mb dew points of 50, 60, 70 and 78°F and bases of 950, 900, 800, and 700 mb, can be computed and are given in Table 15.3. In each case the water released is compared with the amount of water released from a similar cell at 1000 mb and the same temperature. This gives the proportional depletion which results from raising the base and is shown in Fig. 15.9.

Table 15.3. *Effect of raising base of equally proportioned cell*

Pressure at base	Dew Point 1000 mb °F	Released W_e (in.)	Total W (in.)	Released / Released at 1000 mb	Total / Total at 1000 mb
1000 mb	50	0·51	0·84		
	60	0·81	1·38		
	70	1·29	2·27		
	78	1·90	3·35		
950 mb	50	0·42	0·69	0·830	0·825
	60	0·69	1·17	0·856	0·847
	70	1·13	1·96	0·881	0·866
	78	1·70	2·95	0·895	0·880
900 mb	50	0·34	0·56	0·677	0·669
	60	0·58	0·98	0·723	0·706
	70	0·99	1·68	0·766	0·741
	78	1·51	2·57	0·795	0·768
800 mb	50	0·21	0·34	0·417	0·410
	60	0·39	0·64	0·482	0·463
	70	0·71	1·17	0·552	0·518
	78	1·14	1·87	0·603	0·559
700 mb	50	0·11	0·19	0·224	0·223
	60	0·23	0·38	0·290	0·274
	70	0·47	0·76	0·365	0·333
	78	0·82	1·28	0·431	0·382

The precipitable water, W, for the same range of pressures is also computed and shown in Table 15.3, and compared with the total precipitable water in the cell from 1000 mb to the limit of convection at that temperature. The ratios of released water and the ratios of total water do not agree as closely as before when cells of different proportions were considered, but the error is not considered large. Figure 15.9 is used as a measure of the proportional depletion of the released water or the total precipitable water when the base of a convective cell is raised.

Tests of cells of different proportions say 8 : 2 : 8, 8 : 4 : 6, 12 : 0 : 6, etc. may be made, but show, as did the previous table, that, although the amount of water extracted, or the efficiency varies with the models selected, the ratio of moisture released by the model with raised base to moisture released by the same model based at 1000 mb, is similar for all types of models. This means that a storm may be adjusted for elevation irrespective of its actual proportions and the basis of adjustment is the remaining precipitable water.

15.13 Moisture barriers. Topographical barriers may exist between the moisture sources and the project catchments, and cause depletion, intensification, and spillover. The depletion effect is our immediate concern, although intensification and spillover can be dealt with using specific information on storms and catchment.

The height of these barriers to inflow, or moisture barriers, are obtained by averaging the topographical profile at right angles to the inflowing air streams. The height of the moisture barrier depends upon the inflow direction and has a value for each direction of the significant wind streams. Isolated peaks divert the wind flow to either side, and are not as effective as a ridge in producing lift or depletion. The mean height of the range at right angles to the incident stream is a reasonable estimate of the moisture barrier as the higher amounts of moisture passing through gaps below the mean elevation are balanced by the greater depletion caused by the land above.

Rugged and irregular ranges, with differing heights, slopes, and aspects, introduce varied lifts and precipitation rates. This necessitates a smoothing process which generalises or simplifies the topography by cross sections, along the direction of inflow, while profiles at right angles to the inflow permit the estimation of moisture barrier heights.

The barrier reduces the amount of precipitable water which can flow into the project basin. The reduction is equal to the precipitable water in a saturated pseudo adiabatic column of height equal to that of the barrier. This means that the precipitable water flowing into a basin is equal to the precipitable water at the 1000 mb dew point, or at the source region, extending to the limit of convection for that dew point minus the precipitable water from 1000 mb, or source region, to the height of the barrier.

This reduction is approximately 1% per 100 ft, or may be more precisely evaluated from Fig. 15.9 which, although compiled from the proportional depletion with height of the released or effective precipitable water, can also be applied to the proportional depletion with height of the total precipitable water. For example, if an air mass with surface 1000 mb dew point of 70°F passes over a 4000 ft barrier, the depletion is estimated as follows:

(*a*) *Method using basic moisture content or Fig. 13.2*

Precipitable water from 1000 mb to 150 mb at T_d = 70°F = 2·27 in.

Precipitable water from 1000 mb to 868 mb
(4000 ft at 70°F) = 0·76 in.

Precipitable water entering above barrier = 1·51 in.

(b) *Method using moisture depletion or Fig. 15.9*

Precipitable water available (1000 mb – 150 mb at T_d = 70°F = 2·27 in.

Depletion effect of 4000 ft at T_d = 70°F is 0·68

Precipitable water entering above barrier = 0·68 × 2·27 in.

= 1·54 in.

(c) *Approximate estimate.* At 1% depletion per 100 ft gives 40% depletion for 4000 ft.

Precipitable water entering is $\frac{60}{100} \times 2{\cdot}27 = 1{\cdot}37$ in.

This approach assumes that the moisture barrier does not alter the structure of a storm which is merely depleted of part of its moisture on the windward side of the barrier. In more complicated cases intensification and spillover occur.

Moisture barriers, particularly if high, may lift the air mass to the extent that partial or complete convective action is initiated. This may remove and deposit part or all of the effective precipitable water on the windward side of the barrier. The small amount of precipitable water which subsequently finds its way into the valley behind the barrier may yield little or no precipitation when acted upon by the processes available within the valley. This is the well-known rain shadow effect, illustrated by the mountain-encircled inland desert valleys. As maximum rainfall is usually required, any convective action is neglected and the simple depletion with height gives the maximum moisture entering the catchment to produce the greater precipitation. Also, the moisture barriers in project catchments are normally comparatively low and deplete the moisture in accord with Fig. 15.9 rather than trigger off convective rain on the barrier's slopes. If necessary, however, the amount of intensification can be estimated for particular storms and known topography. In addition, when the project basin is close to the moisture barrier and spillover is a significant portion of the precipitation, the barrier adjustment may be either neglected, in full or in part, by moving the moisture barrier appropriately downwind, or the spillover computed from the trajectories of the snow and raindrops formed above the sloping barrier and falling outside the project basin.

15.14 Determining the moisture barriers for a continent. From knowledge of the severe storms and the air mass source areas, the paths of air masses into project basins may be drawn. The land contours athwart these trajectories are the moisture barriers and can be marked on maps. They represent the elevation to which the warm moist air masses must rise in order to enter the project basin or the site of the potential rain fall. For each locality, air mass, and trajectory, there is an appropriate barrier. In this way, maps of a continent may be drawn showing the moisture barriers for the significant air mass and trajectories. It is necessary to completely analyse all major storms affecting the continent. Consequently, maps of moisture barriers are available for only a few countries where such analyses have been done. North America is an example of a continent where these maps are available.

In the absence of maps showing moisture barriers, it is necessary to do a local analysis to obtain the details of significant air masses, their trajectories and the moisture barriers en route to the project basin. The necessary physiographic and meteorological data are usually readily available.

As a general rule, moisture barriers are considered as such only if they are below 3–4000 ft because higher ranges may disturb the air flow to the extent that a storm is modified or changed. In other words, it is not recommended that storms be transposed beyond barriers greater than 4000 ft because the dynamics of the storm may be altered by greater lifts. Basins located behind these higher barriers can be analysed on the basis of local data. Air masses and storms occurring *'in situ'* are examined rather than those beyond the high barriers. In point of fact the mountain ridges frequently delineate the boundaries of regions of meteorological homegeneity and storms should only be transposed within these regions and not across the boundaries.

REFERENCES

[1] BROWNING K.A. and LUDLAM F.H., (1962). Airflow in convective storms. *Q. Jl. Roy. Met. Soc.* **88**, 117–135.

[2] U.S. Weather Bureau (1947). Thunderstorm Rainfall. *Hydrometeorological Report No. 5.*

[3] U.S. Weather Bureau (1960). Generalised estimates of probable maximum precipitation of the United States west of 105th meridian for areas to 400 square miles and durations to 24 hr. *Tech paper No. 38.*

OTHER REFERENCES

U.S. Weather Bureau (1937 to date). *Hydrometeorological Reports and Technical Notes*—various.

World Meteorological Organisation (1969). The estimation of maximum floods. *W.M.O. Tech. Note No. 98.*

CHAPTER SIXTEEN

Estimating the probable maximum precipitation using storm models

16.1 Definition of probable maximum precipitation., P.M.P. It is recognised that there is a physical upper limit to the amount of precipitation that can fall over a specified area in a given time.[1] This upper limit has become known as the Probable Maximum Precipitation, P.M.P., and is more precisely defined as, that depth of precipitation which, for a given area and duration, can be reached, but not exceeded under known meteorological conditions.

The American Meteorological Society in Glossary of Meteorology (1959) expresses this as 'The theoretical greatest depth of precipitation for a given duration that is physically possible over a particular drainage area at a certain time of year. In practice this is derived over flat terrain by storm transposition and moisture adjustment to observed storm patterns'. The term maximum possible precipitation is also used for this concept, but in view of the uncertainties and limitation of data and knowledge of the precipitation process, is considered by some as too definite, and P.M.P. is now generally used. For the same reasons, the P.M.P. must be considered as an estimate, with a range of values depending upon the quality of data, technical knowledge and thoroughness of analysis. It follows that as data, knowledge and analyses improve, so the estimates change, with smaller tolerances and a more realistic representation of P.M.P. The terms extreme precipitation rates and limiting precipitation rates are also used with the same definition and implication.

When actual storms are maximised, a definition of P.M.P. in that region is the critical depth-duration-area rainfall relationships which result when actual storms are increased to represent the most effective rain producing combination of meteorological conditions that would probably occur in that region for the particular month, season or year.

16.2 Probable maximum storm. A distinction is made between the probable maximum storm and the probable maximum precipitation. Different types of storms, say, convective, frontal, convergent, or orographic, may produce the P.M.P. for different durations and areas. The P.M.P. therefore, is made up of a number of probable maximum storms according to duration and area considered. The P.M.P. is thus the envelope to the depth-duration and depth-area curves of the probable maximum storms. For example, the P.M.P. for areas less than 100 square miles and durations less than 6 hr is most likely to be derived from a thunderstorm which is then the probable maximum storm for these durations and areas. For larger areas and longer durations, the P.M.P. values are more likely to be obtained from different storm types say frontal and tropical cyclones, which then become the probable maximum storms for those larger areas and longer durations. A number of probable maximum storms of different durations and areas are thus enveloped by the probable maximum precipitation.

16.3 Seasonal probable maximum precipitation. Although probable maximum precipitation implies the limiting value of precipitation, it may be necessary to determine the highest value which is attained in a given month or season. Circumstances requiring seasonal estimates are as follows:

(*a*) The occurrence of major floods in conjunction with snow melt. The combination of the probable maximum precipitation and snow melt in a particular month may exceed the higher P.M.P. at a time when appreciable snow melt does not occur.

(*b*) The occasion when runoff is materially reduced by high infiltration and peak evaporation rates. This is common in summer.

(*c*) The occasions when more storage is available to retain the higher summer flows than at other seasons. Both (*b*) and (*c*) result in lower discharges below the storage, even although P.M.P. may be higher than during another period when these effects are not in evidence.

(*d*) In a developed river valley, there may be multi-purpose water structures which are designed and operated so that storage is at a low level when the more intense storms are expected. At other seasons, climatological conditions make it advisable to maintain greater storage. Thus a relatively smaller storm at this time may produce a larger flood below the structure than a larger storm at another time when less water is held in storage.

(*e*) In some localities, for example California, the probable maximum storm often occurs in the cooler season as, although temperatures and surface moisture are greater in summer, the significant storm mechanisms, especially for long durations and large areas, do not occur then or are rare.

The significant factors in a hydrometeorological study become apparent after a thorough investigation of the valley and project design together with the history of the storms and floods. All hydrological aspects should be considered including storm structure and movement, runoff, infiltration characteristics of the basin, prevailing winds, moisture sources, snow incidence, and snow melt.

One of the difficulties in obtaining seasonal values of P.M.P. is that the number of intense storms in each month or season is usually relatively low. Although they may be supplemented by careful transposition under controlled conditions, it still represents a relatively small sample. It is necessary to consider whether storms from adjacent periods might not have occurred at other times and so be included. Thus, although the emphasis is upon the possible extreme value in each season or month, investigations show whether meteorological theory permits transposition from adjacent time periods. This is an exception to a general rule that in maximising storms on a seasonal or monthly basis, the maximum dew point within 15 days of the storm date is used. The nature of the storm may warrant an extension of this leeway.

The seasonal variation in the P.M.P. may be assessed by a comparison with the variations of the maximum daily rainfalls in each month. The highest daily rainfalls in each month for all stations in a basin over a compatible length of record are averaged and plotted against time of year to give a seasonal curve. However, as extreme values are of prime interest another index is obtained by taking the five highest daily rainfalls which occurred each month in the valley and plotting these either averaged or individually against their dates of occurrence. The curve drawn through them indicates the likely seasonal variation of P.M.P. Other indices can be devised using frequency analyses of rainfalls over various durations including 2, 3, 5 day, weekly and monthly figures. Duration of storm is not normally significant but can be allowed for in these techniques.

16.4 Probable maximum flood. The probable maximum flood can be obtained from the P.M.P. by subtracting loss rates and adding the snow melt with due regard to relationships with time. This is further considered in texts on surface water hydrology under rainfall – runoff relationships.[2] It is obvious that the probable maximum flood depends upon the assumptions made concerning loss rates and snow melt. If seasonal P.M.P. is to be used, it follows that corresponding values of loss rates and snow melt must be used. That is, consistency has to be maintained.

It might be added that there is no point in accumulating maximum values of all the components producing a flood if analysis shows some of these to be extremely unlikely. Here, the term probable is seen correctly used and in its right context. The term possible might also imply accumulating the extremes. This is another reason for giving preference to the term probable maximum rather than maximum possible.

16.5 Methods of estimating P.M.P. There are a number of methods of computing P.M.P., and as many as possible should be used in a particular study. Some demand an immense amount of work, and so limit their application. The various methods, also, are not independent as the variables considered are inter-related. Storm models, for example, reveal the significant factors which can be maximised in actual storm situations and give the bases for storm transpositions. The various methods have developed from studies of specific catchments, primarily by the Hydrometeorological Section of the U.S. Weather Bureau.[3] As each valley was examined, new problems arose and necessitated the re-examination and modification of methods. The later analyses are more sophisticated and give better results than the early methods but the same principles still apply. There have always been different opinions on the validity and extent of application of the methods. Judgements must be made in interpreting data and in reaching estimates of P.M.P. The following comment by Vance A. Myers is relevant[19] 'The practice of Hydrometeorology has not been reduced to a handbook. No one can furnish a set of rules, graphs, and procedures whereby one can proceed step by step and necessarily drive an acceptable estimate of P.M.P. ..... Handbooks work best in solving uniform problems from data that are uniform and ample. None of these three conditions is the rule in probable maximum rainfall estimates – neither problems nor data are uniform, and the data are certainly not ample'.

Some of the methods which have been used and will be discussed are

(*a*) The storm model approach.

(*b*) The maximisation and transposition of actual storms.

(*c*) The use of generalised data or of maximised depth, duration, area data from storms. These are derived from thunderstorms or general storms.

(*d*) The use of empirical formulae determined from maximum depth, duration, area data, or from theory.

(*e*) The use of empirical relationships between the variables in particular valleys.

(*f*) The statistical analyses of extreme rainfalls.

16.6 The storm model approach. The storm model approach gives the basic principles for most maximisation studies. The precipitation process is simplified by the model so that rates of precipitation may be expressed in terms of physical parameters such as surface dew point, heights of cell, inflow, outflow, etc. Implicit in the storm model approach are the equations of continuity of air flow and of moisture. These enable the flow pattern to be expressed in an equation which gives the rainfall as a function of variables which are measurable, of reasonable length of record, and adequately represent the meteorological conditions, both in time and space. The form of the equation is in accord with meteorological theory, and, the constants of proportionality and integration are found from storm records. That is, both theory and empiricism are needed. It is necessary to test the storm model formulae upon observed storms to confirm that reasonable precipitation values are reproduced, or,that the model is a faithful and reliable one. Any necessary modifications should be made and the formula tested again. This is repeated until the formula is found to be satisfactory.

A statistical analysis of the meteorological variables within the equation enables the extreme values of the variables to be estimated and inserted in the equation to give the Probable Maximum Precipitation. In this part of the procedure, the optimum or most effective combination of the various factors in the equation producing the estimate of P.M.P. must be examined and decided upon. A storm model investigation reveals the significant factors in a high rainfall situation so that these factors may be maximised in an actual storm if it is decided to find the extreme value of precipitation that storm could yield. The conditions under which the storm may be moved to another location may also be discovered from storm models.

The storm model approach is successful in determining the P.M.P. over large areas and in quantitative precipitation forecasting. It is not as successful with small areas due, mainly, to difficulties in measuring thc factors with sufficient precision. The principle of continuity is more easily achieved using mean velocities of inflow and persisting dew points over larger areas and longer intervals of time, than with the brief peaks and extremes which occur locally and over short periods. Network density is also usually insufficient to measure localised moisture supply, wind, and convergence.

In chapter 15 the rainfall intensity in a storm model was shown to depend upon a geometrical basin factor, inflow velocity, a moisture factor, and a convergence factor.

$$\text{e.g. } i = K \,.\, v_{12}\, W_{12} \left\{1 - \frac{\Delta p_{12}}{\Delta p_{34}} \cdot \frac{W_{34}}{W_{12}}\right\}$$

where K is the basin constant. For a specific project catchment, the basin factor is constant for the relevant wind direction.

The inflow velocity (v_{12}) is a necessary part of the convergence mechanism of a storm and may be incorporated with it into one overall efficiency or convergence factor. Thus rainfall (i) can be said to be proportional to a moisture factor times an efficiency or overall convergence factor. The moisture factor may be measured by precipitable water, W, or by specific humidity, q, with corresponding values of efficiency. Some efficiency factors for a particular storm model are given in Table 15.1 and apply to a theoretical cell, but without regard to inflow velocities and areal distribution of the precipitation. The efficiency factor takes into account the dimensions of the storm, air movement, and moisture stratification.

An examination of the actual intense storms which take place in saturated air of similar dew points or moisture content show that the heavy rainfall results from high storm efficiency. The meteorological records also show that air movements, upper air temperatures, pressure gradients and other variables often reach their extreme values within the severe storm experienced. It appears that these storms are at their maximum efficiency, and it is not necessary to increase storm efficiency in maximisation. In any case, the elements contributing to efficiency are probably already at their extreme value and there would be no point in increasing them further. For these reasons, when actual storms are to be maximised, the efficiency factor is not changed, but only the moisture factor. A storm model or synthetic storm on the other hand, has no inbuilt efficiency and requires the substitution of the maximum values of the component variables.

The separate components of the storm model are, therefore, further considered and discussed with a view to extrapolating them to their maximum values. Again, it is pointed out that, although storm models are not always used in obtaining P.M.P., it is necessary to study them to resolve the significant factors for maximisation.

16.7 The basin constant. The basin constant is that part of the precipitation equation which depends upon the geometry of the project catchment and its relation to the inflowing moist air. It is convenient to consider the inflow width at right angles to the air and the downflow length parallel to it. Circumscribing rectangles may be drawn to give these measures when the critical inflow directions are known. Synoptic studies of the intense storms which affect the basin enable the trajectories of the high dew point air masses to be traced from their source, over moisture barriers, into and out of the project basin. From these trajectories, the inflow directions are determined and the basin constants computed. The largest basin constant occurs for a given area when the inflow width is greatest or downflow length is least. This is not always an acceptable value from which to obtain maximum rainfall unless it coincides with the direction giving greatest moisture inflow, that is, the product of basin constant, wind velocity and moisture content must be a maximum and each of these is inter-related.

If significant inflow directions are selected and the appropriate basin constants are combined with inflow velocities, and air moisture contents, allowing for barriers, then the maximum value of the combination can be selected as the significant one for P.M.P.

The basin length, parallel to the downflow direction, should be great enough to allow sufficient time and space for the complete precipitation process to occur within the valley. Where the complete precipitation process does not occur, a correction factor may be applied to the equation. This correction factor is empirically determined from the ratio of the rainfall actually received, to the amount which is theoretically obtainable from the model used with the actual storm parameters.

The amount of precipitable water released depends also upon the model selected. For example, an upglide model is used if frontal or topographical rain occurs, and a convergent model if convection or convergence is the rule. A combination of both is possible. It might be noted that the upglide model is usually applied to larger catchments and both spillover and intensification may occur. The convergent model may be applied to smaller catchments.

Where precipitation is produced by orographical uplift, the direction of slope is important, down slope motion inhibiting rain production. Many directions which give down slope motion can therefore be disregarded in determining basin constants. These directions are self evident from the topography of the area and the storms analysed. A simple correlation of rain with wind direction eliminates all irrelevant wind directions and stresses those important for further investigation.

The basin width should be less than the moist tongue of air. As this is usually about 200–300 miles wide, it is better to restrict basin width to this size. Where greater widths are possible, the valley is subdivided, or a percentage adjustment made to the inflowing tongue of moist air to allow for variations of moisture with latitude.

16.8 Inflow velocities and depth of inflow. The amount of air flowing into the storm model depends upon the inflow velocity, v_{12}, throughout the depth of inflow, Δp_{12}, i.e. $v_{12} . \Delta p_{12}$. The direction and magnitude of the wind are required and synoptic studies indicate the significant direction or directions for intense rainfall. Statistical analyses subsequently give the maximum winds from the significant directions. The average wind velocity over the duration of the precipitation needs to be determined and used in the storm model. The variations of average wind velocities with duration are considered later. Wind velocities used also must reasonably represent inflow in depth, width, and over the area through which inflow occurs.

In practice, it is difficult to determine the inflow velocities due to lack of observations of wind in depth and width during storms. The event of radar winds has improved this by giving information on the vertical wind structure of recent storms but, as these are made at set times, and for a very limited network, information is still lacking on the variation of inflow winds with time and area. Surface wind data from a wider network and including elevated stations, is therefore used to give a better approximation to inflow. The more comprehensive and continuous surface wind records at principal Weather Bureau stations are also correlated with upper wind records and maps, so that winds throughout the depth and breadth of inflow may be estimated. The weather maps which are drawn at regular time intervals, and can be made more frequent if desired by using barograph records, enable geostrophic winds or the resultant of the coriolis and pressure forces, and gradient winds, or the resultant between the coriolis, pressure, and centrifugal forces, to be estimated. From radiosonde data, isobaric maps at various heights may also be used and upper winds deduced. Thus, indications of inflow velocities and their variations with time, height, and area may be obtained.

Each project basin merits its own investigation, but the results from major storms in California may be of interest. These showed that in the lower 10,000 ft the gradient wind equalled 0·8 times the geostrophic wind and from 10,000–20,000 ft the average wind equalled the geostrophic. The average wind over the whole column to 20,000 ft was 0·95 times the geostrophic wind. Consequently the geostrophic wind was accepted as a good measure of the inflow velocity in major storms. If other data were lacking in a project basin, this reasonable assumption may be used to estimate the inflow velocity.

The equation of continuity relates inflow and outflow velocities and their respective depths, viz. $v_{12}\Delta p_{12} = v_{34}\Delta p_{34}$. If three of these are known, the other is also determined. Usually the inflow velocity is observed from radar winds or geostrophic winds as discussed above; the proportions of inflow and outflow Δp_{12} and Δp_{34}, or, storm geometry can be assumed for the model used by referring to actual storms or topography. For example, from cloud and other observations, a thunderstorm or convergent cell may be assumed to comprise three equal portions of inflow, vertical rise and outflow. Here inflow velocity v_{12} equals outflow velocity v_{34}. One orographic model shown in Fig. 16.1, in the

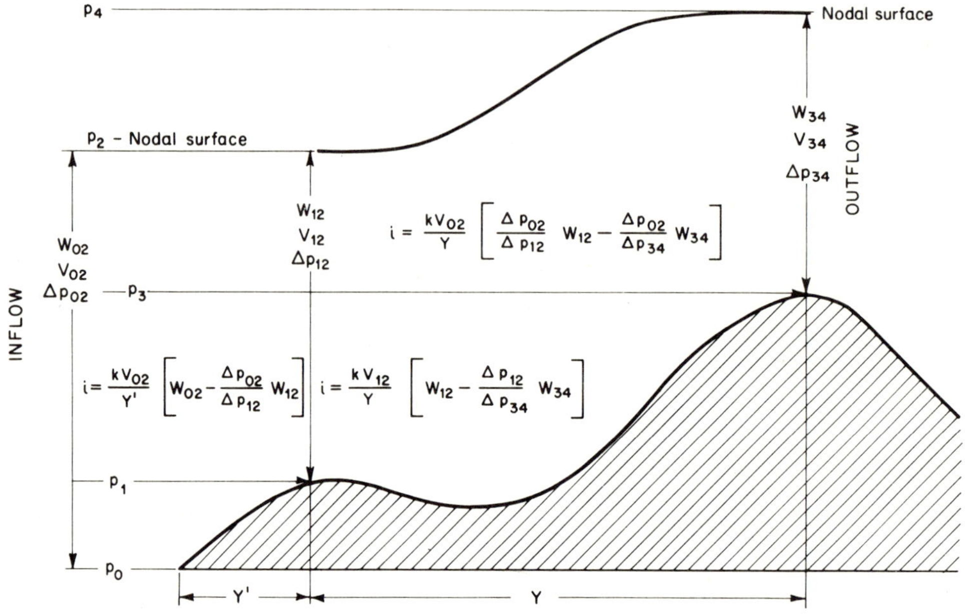

Fig. 16.1. Orographical model in Los Angeles area.

Los Angeles area in the U.S. shows inflow from the surface or 1000 mb to 460 mb and outflow from the top of the range at 684 mb to 300 mb. The upper levels are determined from the wind records of stations located near the inflow and outflow of the basin and represent nodal surfaces. At 460 and 300 mb, or the nodal surfaces, the atmospheric flow is horizontal, showing that the disturbance to flow by topography is below this level. The height of the nodal surfaces or geometry of the model depends upon the nature of the air masses as well as the topography. With known storm geometry, inflow velocity only is required as outflow velocity equals

$$v_{12}\frac{\Delta p_{12}}{\Delta p_{34}},$$

and all these are known.

The average wind velocity for a given duration, which may also be the duration of the storm, is often required and may be obtained from records. In addition, maximum wind velocities over various durations are relevant. It can be shown that logarithms of the maximum wind velocities vary with the logarithms of their duration. By plotting maximum wind velocities, averaged over a given duration against that duration on log–log paper, a straight line is obtained and interpolation for other durations achieved. Weather Bureaux commonly give data on maximum wind gusts and these may be considered as persisting for a half second. The highest average hourly wind and the greatest 24-hr winds are also available. Thus, with due regard for the significant direction of the wind in a storm, the maximum winds may be plotted against their durations on logarithmic paper to obtain interpolations or extrapolations to other durations.

16.9 Maximum winds and P.M.P. When storm models are required to give the probable maximum precipitation, and wind velocity is an essential feature of the storm model, then maximum wind velocities for the significant duration and direction are used.

In major storms it has been observed that pressure gradients, and so the corresponding geostrophic winds, approach, and at time equal the maxima on record for the significant direction and duration. Also, upper air temperatures in these storms are often the lowest on record for that time of year, making the thermal gradients aloft a maximum. These facts suggest that the storms are near maximum efficiency, and involve geometrical proportions and inflow velocities which not only produce heavy rain but, perhaps, may not be exceeded.

As indicated earlier, there are difficulties in determining the wind flow pattern of a storm in depth and over the storm area, as well as its fluctuations with time. The determination of the probable maximum values which these attain can be done as above but with uncertainty. It also follows that the various elements contributing to rainfall intensity are interconnected. For example, if inflow velocities are changed then perhaps surface dew points should be altered due to the more extensive mixing of the air mass at the higher velocities. If maximum wind velocities are used it is logical to use the dew point attained with that wind. This may be considerably lower than the maximum dew point or even the dew point of the storm itself. For these reasons, it is assumed that high intensity storms which have occurred are efficient and already possess an effective combination of inflow velocities and geometrical proportions. In maximisation of actual storms therefore, it is sufficient to increase the water content of the air mass, measured by the precipitable water estimated from the maximum surface dew point. Also, there is the additional advantage that dew points are more easily and reliably evaluated over areas and periods of time than wind velocities. Thus, the intense storms are maximised by allowing for precipitable water only and assuming the storm is of maximum efficiency with optimum inflow wind velocity. Storm models, however, which involve inflow velocities as variables and express the precipitation in terms of inflow velocities, use the maximum values of the wind velocities, consistent with wind direction, moisture charge, and storm duration.

16.10 Moisture adjustment factor. To obtain the P.M.P. the moisture factor is adjusted to a maximum value in storm models and in actual storms. In the case of the storm model, the maximum possible moisture charge for the required duration is determined and substituted in the precipitation equation, together with other appropriate maximum factors, to give the P.M.P. On the other hand, actual storms are raised to a maximum value by using either ratios of effective precipitable water, W_e, or ratios of precipitable water, W, calculated from the 12 or 24-hr persisting surface dew point. That is, the moisture adjustment factor equals either

$$\frac{\text{maximum effective precipitable water}}{\text{actual effective precipitable water}} = \frac{W_e \text{ max}}{W_e \text{ actual}}$$

or the equivalent, for a variety of storm models,

$$\frac{\text{maximum precipitable water}}{\text{actual precipitable water}} = \frac{W \text{ max}}{W \text{ actual}}.$$

Historically, the effective precipitable water was first used as the moisture adjustment factor. It depends not only on the inflowing precipitable water of the air mass, but on the ensuing convergence or storm efficiency, and requires a model to be defined. This is the thunderstorm cell of three equal proportions with a specified upper convective level, which was, firstly, 5 km, then 200 mb, and then a range from 100–300 mb, according to surface dew point. The effective precipitable water is thus given as

$$W_e = \int_{p_1}^{p_2} q\, dp - \int_{p_3}^{p_4} q\, dp,$$

where p_1 to p_2 is inflow and p_3 to p_4 is outflow. The water released depends upon the proportions of the model and as these proportions are varied, different amounts of water are released or W_e varies with the storm model. There is thus little point in using W_e for maximisation and transposition when the storm mechanism is not known. Total precipitable water is therefore preferred for these processes particularly as Tables 15.1 and 15.2 show that the ratio of released water from a variety of cells, for two specific dew points, is equal to the ratio of precipitable water at the same corresponding dew points.

The convergent cell is a highly efficient mechanism for releasing water, and so is often used when there is uncertainty as to the type of process or storm model involved. For example, if orographic uplift causes instability a convergent cell is combined with an orographic model. The orographic uplift produces completely convective rainfall. In this way the greatest amount of water which may be released from the air mass is achieved. If complete convection does not result from the ‘trigger’ action, the estimate is in excess.

The moisture factor is computed in several ways. The *cyclonic adjustment method* assumes that the moisture factor is equal to the precipitable water from the surface, or 1000 mb to 200 mb. Storms are maximised by using the ratios of the precipitable water at the maximum surface dew point to the precipitable water at the representative dew point, assuming a saturated pseudo-adiabatic lapse rate extending to 200 mb. This method is applied to large area storms as it is likely that 200 mb represents the upper limit, or nodal surface, of widespread cyclonic disturbances.

In *the thunderstorm adjustment method* the precipitable water is computed to the assumed upper limit of convection. This is 100 mb at the 1000 mb dew point of 78°F and 300 mb at the 1000 mb dew point of 50°F. Figure 15.7 shows the precipitable water computed to these levels for dew points from 50–78°F. The ratio of the maximum to the actual moisture charge is again the maximising factor. This method is mostly applied to small area and short duration storms where rainfall results from convergent or convective action. The difference between this method and the first is very small as may easily be computed. [4]

The *effective precipitable water adjustment* method was used in early studies and shows only small differences from the thunderstorm or cyclonic adjustment methods which are more convenient to use.

Ratios of *specific humidity* are also used [5] instead of those of precipitable water but not as a common procedure.

The respective adjustment methods vary by less than one percent. In view of the wide availability of Fig. 13.2 and the ease with which precipitable water can be determined, the use of the cyclonic adjustment method is recommended as the most convenient.

When the surface dew point is used for computing precipitable water, the duration of the dew point is relevant. For a storm model, this duration should tally with the duration of the rainfall. However, if actual storms are being adjusted for moisture content the actual period is not important, provided the respective dew points are measured for the same duration. The practice in the U.S. is to use the 12-hr persisting dew point, while in some other countries the 24-hr persisting dew point is more convenient. [6] As explained in chapter 13 the spot, synoptic, or short duration dew point readings are not always representative of the air mass. Therefore, it is an advantage to use dew points of durations exceeding 6 hr, particularly for large area and long duration storms.

16.11 Maximum value of the moisture charge. The moisture charge into a project valley depends upon the basin constant, inflow velocity, and precipitable water of the air mass. For a given basin it is necessary to determine the optimum combination of these factors to give the maximum moisture charge. Analyses of past storms locate sources of moist air and their trajectories into the valley. Moisture barriers en route are taken into account and the basin constants, inflow velocities and dew points are resolved for the significant inflow wind directions or wind circulation. The contribution of each to the moisture charge is computed for all possible cases and the optimum combination of the factors thus obtained.

16.12 Convergence or efficiency in storm models. Only a portion of the moisture charge is released as precipitation. This proportion is a measure of the storm's efficiency or the convergence within the storm.

In the simple upglide model, saturated air of dew point 70°F, if lifted 3000 ft, releases 0·31 in. out of available moisture of 2·27 in. with $0{\cdot}31/2{\cdot}27 \times 100\%$ efficiency (14%). The convergent model in air, of the same dew point, releases 1·29 in. with efficiency $1{\cdot}29/2{\cdot}27 \times 100\%$ (59%). It is assumed that the water vapour is converted to solid or liquid as it reaches saturation, and that the condensation products fall out as precipitation. If the precipitation depth over the valley is required, the basin constant and inflow velocity must be considered as well as the precipitable water and the storm model. When a storm model is set up or defined, the efficiency is decided by its proportions and therefore known. If the storm proportions are changed to give greater efficiency this must be allowed for when the storm model is extrapolated to its maximum value. Normally no change occurs and so no allowance is made for greater efficiency.

16.13 Efficiency of actual storms. Storms occurring within air of a given water content or dew point, display widely differing efficiencies. This is demonstrated in Fig. 16.2 where the precipitation depth for a given area and duration is plotted against the storm's moisture factor, measured by precipitable water. Alternatively specific humidity may be used.[7] The slope of the line joining the plotted point to the origin represents the convergence factor or efficiency of the storm. The many storms show a vast range of efficiencies with relatively few of steep slope or high efficiency.

Depths achieved in some of the major storms in the United states[8] over an area of 5000 square miles and for a duration of 24 hr are shown in Fig. 16.2 and given in Table 16.1, together with the precipitable water of the air mass, the average storm intensity,

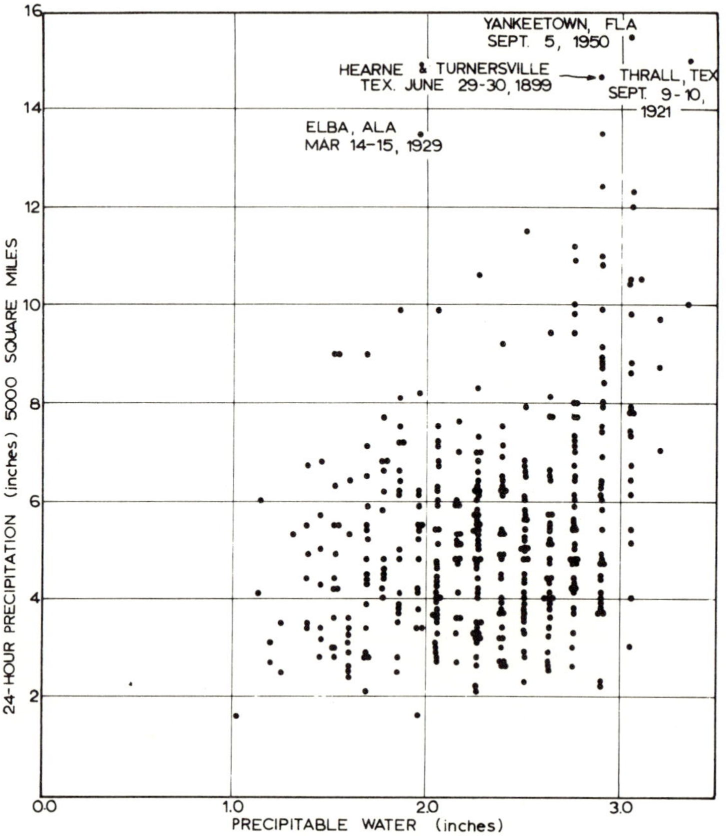

Fig. 16.2. Precipitable water plotted against maximum 24 hr 5000 square miles precipitation for major U.S. storms.

and the relative efficiency, assessed as the intensity divided by precipitable water. It is seen that, although total depth of precipitation increases with precipitable water, high efficiencies occur in air masses of a wide range of precipitable waters. The fact that the efficiency does not vary with the moisture content of the air suggests that a moisture adjustment of the efficient storms does not alter their efficiency.

As interest is in the major storms, the maximum value of the convergence or efficiency factor is obtainable for a given duration and area by an envelopment of all points of the

Table 16.1. *Major storms in U.S. of 24 hr duration over 5000 square miles*

Precipitation depth (in.)	6	9	10	13·5	14·6	15·0	15·5
Average intensity i in./hr	0·25	0·38	0·42	0·56	0·61	0·63	0·65
Precipitable water, W, (12 hr dew pt) in.	1·2	1·5	1·8	2·0	2·9	3·4	3·1
Efficiency (i/W).	0·21	0·25	0·23	0·28	0·21	0·19	0·21

plotted storms. This is shown in Fig. 16.2 with the data of Table 16.1 and other intense storms for 5000 square miles and 24 hr. The maximum efficiency achieved, or the maximum convergence, is

$$6{\cdot}9\,\frac{\text{in. per 24 hr over 5000 square miles for 24 hr}}{\text{precipitable water, } W \text{ in.}} \text{ or } 0{\cdot}29\,\frac{\text{in./hr over 5000 square miles}}{\text{precipitable water, } W \text{ in.}}$$

For other durations and areas, the maximum storm efficiencies may be determined in a similar way. It becomes apparent[7] that as area and durations decrease, the efficiency increases. That is, the most efficient storms occur over small areas and for short periods. The physical structure of storms supports this contention. Also it is likely that the variations in rainfall intensity during a storm and over an area, result from alterations in the storm mechanism. The air mass shows some uniformity of water content in space and time, but inflow and outflow heights and velocities, including vertical velocities, do show considerable fluctuations. These features of the storm model give variations in moisture storage and deposition in space and time. The determined values of storm efficiencies apply to storms occurring over flat ground because topography, by intensification and spillover, alters the efficiency of a storm. The topographical effect is not easy to determine although it is possible to evaluate the influence of the various parameters such as slope, orientation exposure etc. especially for the average case.[9] Topographical storms display a wide range of efficiencies related to their sites and are best used within their particular region or areas of similar topography.

Under normal circumstances, actual storms are not adjusted for efficiency for the following reasons.

(*a*) Precipitation theory is inadequate to fully estimate the storm efficiency or the greatest efficiency which it may reach.

(*b*) The major storms have produced heavy precipitation because of high efficiency rather than high moisture-content. That is, the highly efficient storms have occurred and are contained in the record.

(*c*) Extreme conditions reported in the major storms suggest that they approached maximum efficiency. For example, maximum pressure gradients with minimum central pressure have been recorded in these storms. Also, the amount of precipitable water leaving a storm cell depends upon temperatures aloft, and as these have approached the extreme minima for the period of record in the major storms, it is likely that minimum moisture outflow and so high efficiency has already occurred in the storms studied with no scope for adjustment to higher efficiency.

(*d*) The wide range of precipitation depths associated with the one surface dew point suggests that some involve the most effective combination of meteorological factors and approach maximum efficiency.

(*e*) The moisture adjustment used for actual storms may be an over-adjustment and thus allow for any slight increase in efficiency. Also, higher efficiency may require stronger winds which are probably associated with lower dew points than the winds at the maximum dew point.

16.14 The location of storms of high efficiency. Intense storms occur in preferred areas because of the presence of high dew point air mass sources nearby, and the movement of air from these sources into the preferred areas. Storm mechanisms need to be efficient and may occur relatively frequently in the area. The most efficient mechanism is that of convergence over small areas and short duration. Intense storms of this type, akin to the thunderstorm, occur over a wide range of latitudes but most frequently, and with highest intensity, in the tropical and subtropical zones where the incursions of tropical and equatorial air are common. The intense storms over larger areas and durations usually involve several mechanisms, but convergence, typified in the tropical cyclone or trough, again gives high rates of rainfall. The optimum combination of significant factors occurs, in the United States and Australia, around about latitude 30 degrees and on the eastern sides of the continents. Here there are high dew points, unstable air masses, and frequent basic storm mechanisms. The efficiency of the major storms correlates highly with their moisture content. It is doubtful whether the efficiency factors of these storms are applicable beyond their meteorologically homogeneous zones, or, say, to the western coasts. However, when there is a short history of storms within an area, it is necessary to examine the efficiencies of all storms to see if any could reasonably be transposed. Efficiency is determined from the relationship between precipitation depths and moisture content for specific storm areas and durations.[7] Efficiencies may be determined for

(*a*) the immediate project area,
(*b*) the meteorologically homogeneous zone,
(*c*) the similar type storms from other zones, and
(*d*) all storms irrespective of type and location.

Storm analyses and the efficiency factors obtained indicate whether the storm history of the area contains efficient storms and the degree of extrapolation which is permissable to the efficiency factors in the project area. Instead of estimating efficiency factors, a more liberal transposition of actual storms may be preferred.

16.15 Examples of storm models. Storm models are used to estimate the P.M.P. when there is insufficient or un-representative storm data, rugged topography, or a degree of rationalisation and consistency is needed.

The storm data may be deficient, in that few efficient storms occurred within the area, or that the instrumental network was inadequate or faulty when the efficient storms occurred. A network is not representative when there are difficulties in finding correct exposures for rain gauges.

Varied topography gives a diverse rainfall pattern and presents problems in selecting or setting up a rain gauge network representative of the whole catchment. Besides this difficulty in siting rain gauges, rugged topography leads to varied depletion, intensification and spillover effects with complicated isohyetal patterns. Transposition of storms from other catchments under these conditions can be inaccurate. Again, in rugged areas it frequently happens that there are only a few storms '*in situ*' and these are poorly documented. Thus the storm model approach may be the only one left for estimation of P.M.P.

As regards consistency, the data on the basin may give a wide scatter of values of P.M.P. and a storm model may enable a rational basis to be given to these estimates, minimise inconsistencies, and show potential sources of error.

The storm models used are based on the processes of upglide and convergence which have already been discussed. The upglide model is typified by orographical rainfall while examples of the convergent process are the thunderstorm and the tropical cyclone. More specific details of these storm models are now briefly considered.

16.16 Orographical models. The orographical models[10,11] shown in Figs. 16.1 and 16.3 give fixed conditions of height of the inflow barrier, p_1, and the main range, p_3.

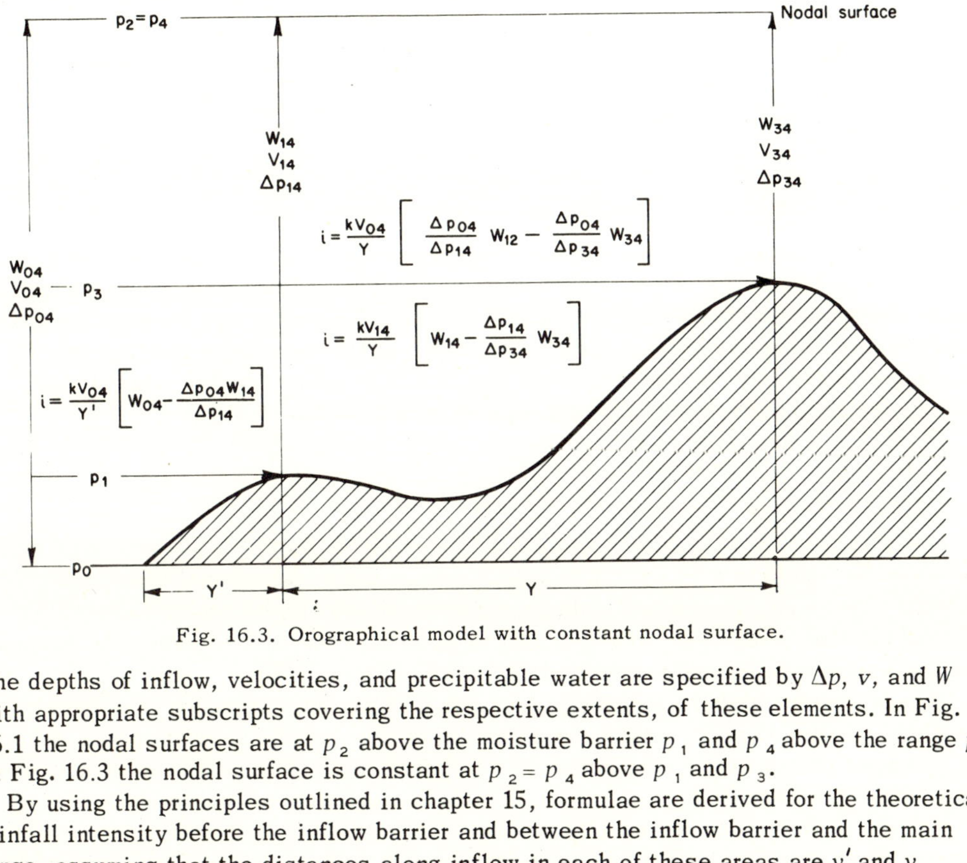

Fig. 16.3. Orographical model with constant nodal surface.

The depths of inflow, velocities, and precipitable water are specified by Δp, v, and W with appropriate subscripts covering the respective extents, of these elements. In Fig. 16.1 the nodal surfaces are at p_2 above the moisture barrier p_1 and p_4 above the range p_3. In Fig. 16.3 the nodal surface is constant at $p_2 = p_4$ above p_1 and p_3.

By using the principles outlined in chapter 15, formulae are derived for the theoretical rainfall intensity before the inflow barrier and between the inflow barrier and the main range, assuming that the distances along inflow in each of these areas are y' and y respectively. The proportionality factor K allows for the geometry of the catchment. The formulae given in Figs. 16.1 and 16.3 give the precipitable water released by the assumed lift at the barriers. The topography is generalised both along the inflow direction and at right angles. The latter is used to give the inflow barrier and the height of lift by the main range.

The information available makes it possible to draw streamlines of air and moisture flow. These are shown diagramatically in Fig. 16.4 where the three processes of depletion, intensification, and spillover are apparent. Depletion occurs before the inflow

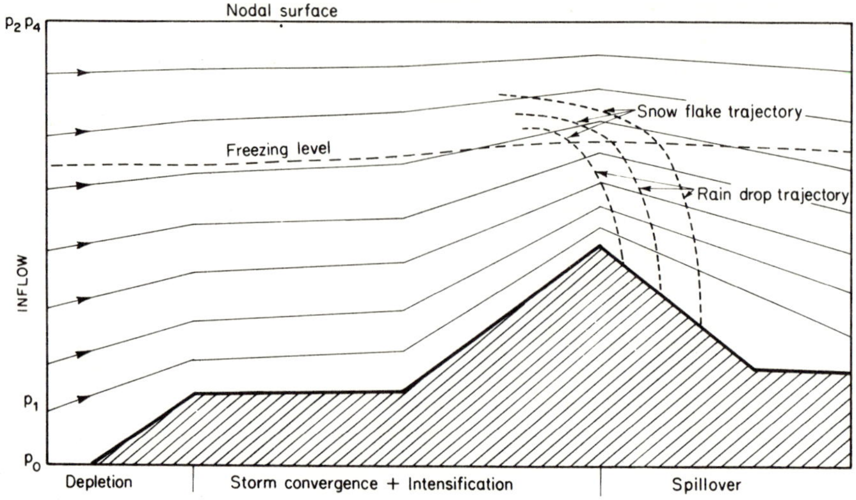

Fig. 16.4. Streamlines of air flow and precipitation trajectories in a storm model.

barrier and the available precipitable water is reduced by the liquid equivalent in a column of air equal to the height of the barrier. Although spillover may occur, in practice this is small as inflow barriers are comparatively low. Allowance may be made for spill-over if desired. Intensification occurs on the slopes of the main range and can be measured by continuity considerations and the wind flow defined by the streamlines. The intensification effect can also be computed from equation 14.2, viz.

$$i = \frac{w_c \rho_c (r_c - r_z)}{7},$$

and assuming the inflowing wind v against the orographical barrier which makes an angle α with the horizontal to give a vertical velocity $w = v \tan \alpha$. As vertical velocities decrease with height a refinement reduces the vertical velocity by one half every 1000 m.[12] Also geostrophic winds and actual wind data are used to define the stream-line and maximum velocities for the various periods of time.[13]

If intensification leads to convection as a result of the latent instability of the air mass, this is allowed for by raising the levels of outflow. In this case the mountain range has a 'trigger' effect upon the air mass. The height of convection is dependent upon the properties of the air mass.

Spillover, or the falling of raindrops or snow developed at height above an area into an adjacent catchment, can be computed if sufficient data is to hand.[13] The nature of the air mass, height of the freezing level, its horizontal and vertical velocity structure, the size of drops formed, and their velocity of fall, are used to draw streamlines of air flow and the trajectories of the snow and rain. By relating these to the topography the amount of spillover is computed.[12,13]

The formulae established by the above reasoning are tested by applying them to intense storms within the area and modifications made if necessary. Extrapolation to the probable maximum values of the component elements of the formulae gives an estimate of probable maximum precipitation. The use of a computer makes the work of estimation less tedious and enable a number of trials to be made.[12]

16.17 Convergent models. In chapter 15, convergence is discussed and Figs. 15.6 and 15.7 show the ideal convergent cell with inflow in the lower layers and outflow aloft. The thunderstorm is a typical example of this model. In practice the air flow is more complicated, as can be seen from Fig. 14.1 which is the likely air flow in a typical hail storm. This example shows that strong localised vertical currents are necessary to develop and support large drops and hailstones in some areas, but the lower velocities and downdrafts in other sections enable the descent of the precipitation products. The overall flow outside the central section is one of convergence in the lower levels and divergence aloft. The cloud structures of the hail storm and thunderstorm show this distinctive pattern with anvil cirrus in the divergent upper section. Estimates of extreme precipitation can be made by applying this convergent cell to a basin and using the principles of chapter 15. Horizontal movement may also be imposed on the cell. In actual practice, the ideal structure does not exist, the convergent cell has a short life over a small area and is translated horizontally, accurate measurements of the elements concerned in both the short periods and areas concerned are not available and water storage in the cloud is not considered because of computational difficulties, although it is probably high because of strong vertical currents. Thus the method is not a practical one for estimation of P.M.P. unless the very high values of the ideal cell are used as an overall extreme value. It is preferred to use the depth, duration, area values of the small area, short duration storms and the cloud bursts falls given in chapter 17.

The tropical cyclone is a typical example of the convergent model which has been used to estimate its P.M.P.[4] The horizontal wind pattern defined by Graham and Nunn[15] and the vertical structure given by Riehl[16] and Miller[17] form the basis for the model approach to the tropical cyclone. Figures 16.5 and 16.6 show a likely horizontal wind field and vertical structure of a tropical cyclone in the northern hemisphere. A mirror image represents the case in the south.

In Fig. 16.5 the wind velocities are indicated by the isopleths and the directions by the arrows. The wind makes an angle of 25° with the tangents to the circles drawn about the wind centre. The radial velocity may thus be computed. Although conditions are represented at 30 ft above a water surface for a large hurricane moving about 10–15 miles/hr, the known variation over land and with height suggest no great error in using Fig. 16.5 to represent the lower 1 km inflow layer.

In Fig. 16.6 there is inflow in the lower 1 km or 100 mb layer, no convergence or divergence between 1 and 6 km and divergence or outflow above 6 km. If this is accepted as the hurricane model, suitable formulae may be derived. For example, if air of specific humidity $\bar{q}$ flows into a tropical cyclone cell of radius r with radial velocity v_r, it was shown in chapter 15 that the rainfall in unit time and area was

$$\frac{2v_r}{r} \cdot \bar{q}\,\frac{p_{o1}}{g}$$

where g is the gravitational acceleration.

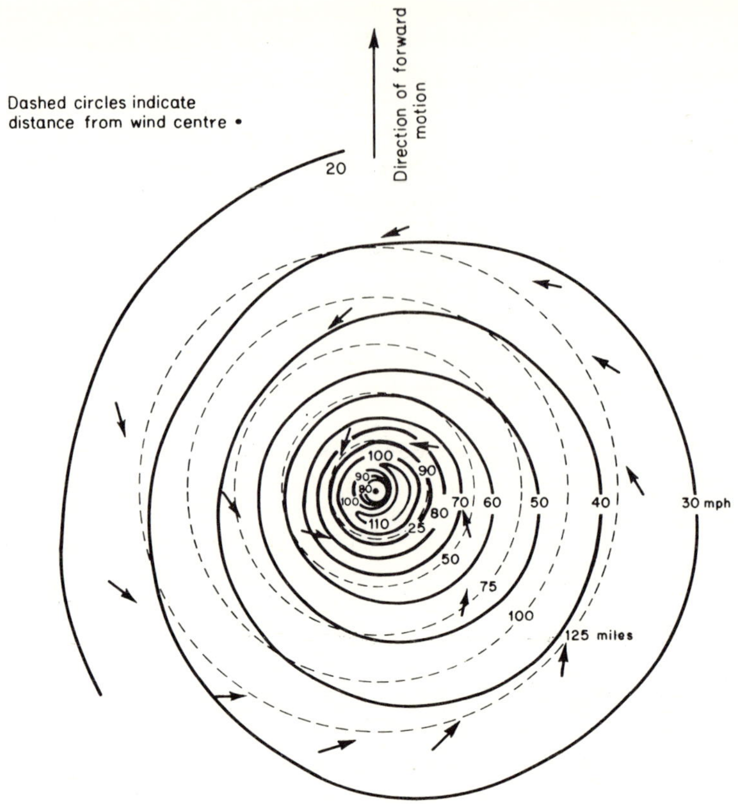

Fig. 16.5. The surface wind field in a tropical cyclone.

If $\bar{v}_r$ is the mean radial inflow velocity in m/sec and $\bar{w}$ the mean mixing ratio in g/kg and r the radius in miles, then rainfall intensity in./hr is

$$\frac{0{\cdot}18\bar{v}_r\bar{w}}{r}$$

assuming negligible outflow.

If the inflow structure of Fig. 16.5 is assumed, and a base temperature of 75°F and a saturated pseudo-adiabatic lapse rate, the rainfall intensities at various distances from the centre can be computed and are given in Table 16.1.

A movement may be assumed for the model (5 miles/hr) and estimates obtained for the probable maximum precipitation.

The hurricane model can be combined with orographical uplift to give the probable maximum precipitation on an elevated ridge in its course.

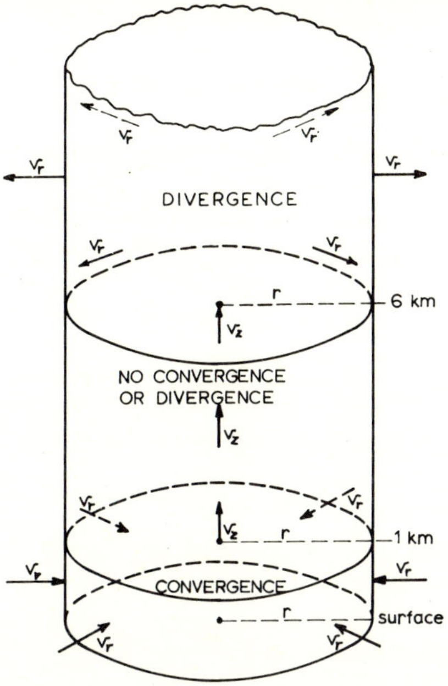

Fig. 16.6. The vertical structure of a tropical cyclone.

Table 16.1. *Rainfall intensities at various distances* (r) *from a hurricane centre.*

r (miles)	V_r (m/sec)	i in./hr within circle of radius r	i in./hr within adjoining circles $r_2 - r_1$, etc
5	0	0	9·25
10	20·8	6·92	2·36
20	21·0	3·50	0·73
30	17·7	1·96	0·39
40	15·0	1·25	0·21
50	13·1	0·87	0·13
60	11·5	0·64	0·10
70	10·6	0·50	0·07
80	9·6	0·40	0·05
90	9·0	0·33	0·03
100	8·3	0·28	0·01
110	7·9	0·24	
120	7·3	0·20	

REFERENCES

[1] BERNARD M., (1944). The primary role of Meteorology in flood flow estimating. *Trans ASCE* **109**, 311–382.

[2] CHOW V.T., (1964). *Handbook of Applied Hydrology*, McGraw Hill, New York.

[3] U.S. Weather Bureau (1937 to date). *Hydrometeorological Reports* and *Technical Notes*. U.S. Govt. Printing Office.

[4] PAULHAUS J.L. and GILMAN C.S., (1953). Evaluation of probable maximum precipitation. *Trans. Am. Geophy. Uni.* **34**, 701–708.

[5] FLETCHER R.D., (1951). Hydrometeorology in the U.S. *Compendium of Meteorology*. Am. Met. Soc.

[6] WALPOLE J., (1958). Maximum possible rainfall over Australia. *Conference on Estimation of extreme precipitation*. Bureau of Meteorology, Melbourne, Australia.

[7] FLETCHER R.D., (1948). Computation of thunderstorm rainfall. *Trans. Am. Geophys. Uni.* **29**, 41–50.

[8] U.S. Weather Bureau (1960). Generalised estimates of probable maximum precipitation for the United States west of the 105th meridian. *Tech Paper No.* 38. U.S. Govt. Printing Office.

[9] SPREEN W.C., (1947). Determination of the effect of topography upon precipitation. *Trans. Am. Geophys. Uni.* **28**, 285–290.

[10] U.S. Weather Bureau (1945). Revised report on maximum possible precipitation Los Angeles area, California. *Hydromet. Report No. 21B*.

[11] U.S. Weather Bureau (1947). Maximum possible precipitation. San Joaquin Basin, California. *Hydromet Report No.* 24.

[12] KNOX J.B., (1960). Procedures for estimating maximum possible precipitation. *Bull No.* 88 dept. of Water Resources, State of California.

[13] U.S. Weather Bureau (1961). Interim Report, Probable maximum precipitation in California. *Hydromet. Report No.* 36.

[14] U.S. Weather Bureau (1961). Generalised estimates of P.M.P. and rainfall frequency data for Puerto Rico and Virgin Islands. *Tech. Paper No.* 42.

[15] GRAHAM H.E. and NUNN D.E., (1959). Meteorological considerations pertinent to standard project hurricane, Atlantic and Gulf Coasts, *National Hurricane Research Project Report No.* 33. U.S.W.B.

[16] RIEHL H., (1954). *Tropical Meteorology*, McGraw Hill, New York.

[17] MILLER B.T., (1958). The three dimensional wind structure around a tropical cyclone. National Hurricane Research *Project Report No.* 15 U.S.W.B.

[18] U.S. Weather Bureau (1969). Probable maximum precipitation over South Platte River, Colorado and Minnesota River, Minnesota, *Hydromet. Report No.* 44.

[19] United Nations E.C.A.F.E. and W.M.O. (1967). Assessment of the magnitude and frequency of flood flows, *Water Resources Series No.* 30.

CHAPTER SEVENTEEN

Estimating the probable maximum precipitation — Actual storms and transposition

17.1 The maximisation of actual storms. In discussing the maximisation of storm models, it should now be clear that data and theories on air masses, storm structure, and precipitation are limited and inadequate. Therefore and in view of the many assumptions and generalisations made, it is not surprising that estimations of probable maximum precipitation from models have aroused controversy. [1,2] The criticism made is that the models oversimplify and do not fully represent the many variables and their interrelations. It is claimed they give unrealistic values of P.M.P. In reply it may be stated that models reveal the significant factors for maximisation, and where data is limited, may offer the only solution. Also, storm models give reasonable results when large areas and long storm durations are considered as the broad generalisations and use of the principles of continuity hold for extension in time and space. Where the lifting mechanism can be faithfully reproduced by a model as in topographical storms, worthwhile solutions are obtained. However, over small areas and for short durations, there are many fluctuations in the storm's convergence and lift as well as in the moisture inflow, and therefore it is not possible to represent these effects suitably in a storm model. Successful storm models have been applied to basins where large areas, long durations, and topographical features are significant.

Due to the deficiencies of the storm model approach and to give additional evidence, actual storms have been extensively studied and extrapolated to their probable maximum values. Together with the principle of transposing storms from other areas, this method constitutes the most convenient and acceptable form of obtaining the probable maximum precipitation. It involves the collection of meteorological data for the extreme storms which have occurred within the project basin and their analyses, including the drawing of appropriate isohyetal maps, construction of mass curves, and the estimation of the moisture charge from the representative surface dew points of the storms. The storm rainfall depths obtained either as isohyetal maps or D.D.A. values, are used with the storm's representative dew point and the maximum dew point which can occur in that region near the season of the storm to give the probable maximum precipitation for the basin. This approach avoids many of the difficulties of the storm models, topographical intensification, spillover, and variable storm efficiencies. Its success depends on whether sufficient severe storms, adequately described by data, have occurred within the project valley.

Storm models show that the rainfall intensity is proportional to a moisture factor and an overall convergence factor. If the moisture factor is expressed in terms of precipitable water, then actual storms may be adjusted, using (*a*) The cyclone model with W evaluated to 200 mb.

(*b*) The thunderstorm model with W evaluated to the convective level.

(*c*) Specific humidities, but not as a common practice.

Although various storm models release different amounts of moisture, it was established in chapter 15 that the ratio of released water at two separate dew points is the same as the ratio of the total precipitable water at the same two dew points, irrespective of the model. It is more convenient to use the total precipitable water and adjust an actual storm to its probable maximum by using the ratio of

$$\frac{\text{precipitable water at the maximum persisting dew point of the zone}}{\text{precipitable water at the representative persisting dew point of the storm.}}$$

Actual storms are not adjusted for efficiency, as it is assumed that, amongst the major storms, some approach maximum efficiency for the zone and season. Arguments supporting this contention are given in paragraphs 16.6 and 16.13.

Precipitation and dew points vary with area and duration. Accordingly, it is necessary to use isohyetal maps or depth, duration, area data for storms and determine a representative dew point persisting over a period, usually 12 or 24 hr. If relevant, seasonal data is used and adjusted to the maximum dew point occurring within 15 days of the actual storm, unless analysis indicates wider limits. Provision of data is a function of the national meteorological service which is sometimes assisted by other authorities. There is a wide range of publications. Also internal files and original records are usually made available. If necessary, the appropriate analyses may be done from basic data using the accepted and approved methods described in texts and publications.[3–5]

17.2 Maximising storms '*in situ*'. Often, there are local influences on storms which show up many variations in rainfall intensities and complicated isohyetal patterns. Usually this occurs in mountainous regions where project basins are commonly located. Because of these local influences, estimations of probable maximum precipitation are most successful if they are based on the intense storms which have occurred within the valley and are maximised using the highest recorded dew point of that valley. If relevant, seasonal figures, are considered. Maximisation of the storm '*in situ*' allows for topographical influences, spillover, or other local effects. These effects can be large and difficult to assess but are reflected in the local storm and its maximised counterpart.

The drawback to this approach is a lack of data on intense storms within valleys. It may be that such storms have not occurred, but it is more likely that they did but were not recorded. Sometimes it is possible to obtain information from private gauges, newspaper reports, or local inhabitants. This emphasises the necessity for the early setting up of networks, or the taking of special surveys after severe storms to collect data not normally available. Information from containers left outside, erosion effects, or flood marks are examples of useful data collection.

Storms maximised '*in situ*' can be supplemented by transposed storms. These may indicate seasonal variations in rainfall intensity, confirm whether the valley storms approached maximum efficiency, or indicate if more severe storms might not occur. Greatest weight should be given, however, to the maximised '*in situ*' storms.

17.3 The transposition of storms. Within a specific river valley there are often inadequate records of storms. In any case, it is good practice to transpose all outstanding storms from other areas to the project basin if it is at all likely that they could have occurred there. This possibility depends upon meteorological factors, chiefly inflow velocities and storm structure. If these are unaffected by transfer, the storm may be transposed. The expression of storm precipitation in terms of a moisture and a convergence factor assumes that, in the zone of transposibility, the convergence factor is not significantly changed. In transposition therefore, the moisture factor only is adjusted.

17.4 Meteorologically homogeneous zones. Zones of meteorological homogeneity are defined as areas where every point can experience a storm event with the same storm mechanism and total inflow wind movement but not necessarily with the same moisture charge or with the same frequency. It follows that storms may be transposed within the meteorologically homogeneous zones.

These zones are determined principally by subjective considerations using synoptic maps, aerological data, and depth, duration, area information. When a large number of storms have been analysed, their locations and similarity of features suggest certain homogeneous zones. In addition, meteorological theory imposes limitations on transfer. For example, tropical cyclones cannot be transposed to polar latitudes without profound changes as conditions for their existence are limited to tropical, and subtropical latitudes. Usually zones do not encompass more than 10° latitude for this reason.

The topography of a region may affect moisture flow and the storm structure to the extent that a cyclone may not move into a region behind a high barrier, without fundamental changes. Frontal surfaces can also be destroyed or modified by their passage over rugged terrain. Consequently high mountain ranges usually form the boundaries of meteorologically homogeneous zones rather than intersect them. Transposition of storms in the mountains themselves presents a special problem to be discussed later.

The location and extent of air masses of appropriate properties also determine the zones of meteorological homogeneity. Certain storms require the presence of active moist air masses which might only be found in special coastal regions. Similar storms may not be transposed outside the known extent of the air masses.

The determination of the zones of meteorological homogeneity is thus a complex problem which can be solved after detailed storm analyses. This has only been done in a few places. In the United States approximately 21 sub-divisions are delineated but some storms can be transposed into more than one subdivision. The zones are a guide only and the limits of transposability of each storm should be decided after considering these zones and the storm itself. In particular, the storm's surface and upper air properties, its location with respect to topography and the general circulation, and the factors within it causing the heavy rain, must be considered. The degree to which these factors approach their maximum possible value is significant as well as the dependence of the storm upon local conditions which may not be available at other sites. Orographical storms and local severe squalls and cloud bursts are examples of intense storms which are enhanced or induced by topography and may not be transposed to different sites without modification.

17.5 Basic assumptions of storm transposition. In moving a storm to a new location it is assumed that no change is made in its moisture stratification, its spatial dimensions, and in the magnitude and distribution of the inflow and outflow winds. It might be noted that the process of maximisation has similar assumptions of no change in these factors with increased moisture inflow. Although our knowledge of vertical wind structure and its variations with topography and site as well as variations in storm dimensions is limited, this assumption is considered reasonable.

The earlier assumption that the surface dew point and the saturated pseudo-adiabatic lapse rate represents the moisture content of the inflowing air mass is also required. The slight over-estimate which occurs with this assumption may compensate for the occurrence of a more efficient storm for which no adjustment factor is used.

In transposition, a larger number of major storms are made available to the project basin and there is a greater chance that at least some have approached the optimum combination of inflow wind and storm mechanism or are of maximum efficiency. Thus it is assumed that most meteorologically homogeneous zones have some of these high efficiency storms and their maximisation and transposition within the zone gives the probable maximum precipitation. In addition, storm models, generalised estimates, data from other zones and empirical formulae give check estimates.

Another assumption is that the same adjustment factor can be applied to storms of various durations. For reasons already given, the surface dew point, which persists for 12 for 24 hr, is used for both the representative storm dew point and the maximum dew point. Records of dew point show a decrease of dew point with increased duration but this applies to both the dew points during storms and the maximum dew points encountered. The duration graphs of these are lines slightly converging with duration indicating that the ratios of maximum precipitable water to actual storm precipitable water would be reasonably constant for dew points of all durations.

17.6 Transposing storms. In transposing storms to a project basin the following procedure is carried out

(*a*) Select storms of the appropriate season from the meteorologically homogeneous zone for transposition. Depth, duration values of the storm for the area of the project basin are required together with the representative dew point of the storm, usually given at 1000 mb.

(*b*) Determine the inflow barriers for the project basin and the storm under transposition. The trajectories of the air masses concerned must be considered. The average height of the basin may differ from the height of the moisture barrier in the path of the incoming air. The usual procedure is to use the height of the moisture barrier. In most cases there is only a small difference.

(*c*) Determine the maximum dew point which can occur within the basin and express this at 1000 mb.

(*d*) Obtain the appropriate adjustment factor or factors. This may be done in one stage by determining the precipitable water at the site of the actual storm, so allowing for the storm's moisture barrier, and dividing it into the maximum precipitable water at the new site depleted by the new moisture barrier. Alternatively, three adjustments may be made for

(*i*) Maximum dew point at the site of the storm.

(*ii*) Maximum dew point at the project basin.

(*iii*) Difference in elevation between the storm's site and the transposed site.

Ratios of precipitable waters are used and the three factors are combined, i.e. (*i*), (*ii*) and (*iii*). This becomes the same adjustment factor as the single stage approach due to terms in the numerator and denominator cancelling out.

Example of storm transposition. A storm has observed depth of 8 in. over 1000 square miles for 6 hr; 10 in. for 12 hr; 12 in. for 18 hr; and 13 in. for 24 hr. It occurred with an inflow barrier of 1000 ft and a 1000 mb dew point of 70°F. This storm is to be transposed to a catchment of 1000 square miles in the same meteorologically homogeneous zone with a moisture barrier of 1500 ft. The highest 1000 mb dew point during this season was 75°F at the transposed site and 76°F at the actual site.

Method 1. Single adjustment factor.

Observed storm

Total W from 1000 mb to 200 mb at 70°F	=	2·27
Depletion W from 1000 mb to 1000 ft at 70°F	=	0·21
Residual W from 1000 ft to 200 mb at 70°F	=	2·06

Project basin

Total W from 1000 mb to 200 mb at 75°F	=	2·90
Depletion W from 1000 mb to 1500 ft at 75°F	=	0·37
Residual W from 1500 ft to 200 mb at 75°F	=	2·53

$$\text{Adjustment factor} = \frac{\text{Residual } W \text{ project basin}}{\text{Residual } W \text{ observed storm}}$$

$$= \frac{W \text{ from 1500 ft to 200 mb at 75°F}}{W \text{ from 1000 ft to 200 mb at 70°F}}$$

$$= \frac{2{\cdot}53}{2{\cdot}06} = 1{\cdot}23.$$

Adjusted depths for the basin or P.M.P. are 6 hr, 9·8 in.; 12 hr, 12·3 in.; 18 hr, 14·8 in.; 24 hr, 16·0 in.

Method 2. Separate adjustments for dew points and height.

(*1*) Adjustment for $M \times$ dew pt at storm
$$\frac{W \text{ from 1000 ft to 200 mb at 76°F}}{W \text{ from 1000 ft to 200 mb at 70°F}} = \frac{2{\cdot}78}{2{\cdot}06}$$

(*2*) Adjustment for $M \times$ dew pt at project
$$\frac{W \text{ from 1000 ft to 200 mb at 75°F}}{W \text{ from 1000 ft to 200 mb at 76°F}} = \frac{2{\cdot}65}{2{\cdot}78}$$

(*3*) Adjustment for height difference of sites
$$\frac{W \text{ from 1500 ft to 200 mb at 75°F}}{W \text{ from 1000 ft to 200 mb at 75°F}} = \frac{2{\cdot}53}{2{\cdot}65}$$

$$\text{Total adjustment} = (1) \times (2) \times (3) = \frac{2{\cdot}78}{2{\cdot}06} \times \frac{2{\cdot}65}{2{\cdot}78} \times \frac{2{\cdot}53}{2{\cdot}65}$$

$$= \frac{W \text{ from 1500 ft to 200 mb at 75°F}}{W \text{ from 1000 ft to 200 mb at 70°F}} = \frac{2{\cdot}53}{2{\cdot}06} = 1{\cdot}23.$$

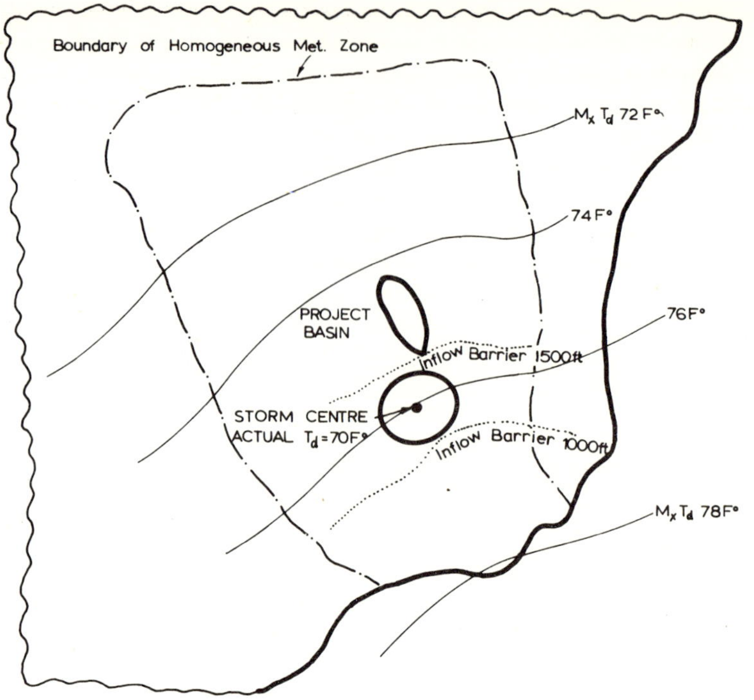

Fig. 17.1. Transposition and maximisation of a storm.

Figure 17.1 is a diagrammatic representation of the transposition and maximisation of the above storm. Isopleths of maximum dew point are shown for the homogeneous meteorological zone within which the transposition takes place. The moisture barriers are given for the significant wind direction of the transposed storm.

17.7 Storm orientation. Most storms show a comparatively regular pattern of rainfall, especially if there are no topographical influences. Others give irregular or elongated patterns which are related to the storm mechanism or moisture inflow rather than to topography. Such storms may be transposed, but the orientation must be examined to see if this is a necessary feature of the storm. For example, the fronts of an extratropical cyclone may be linked to a specific isohyetal pattern which is parallel to the fronts concerned. A study of similar cyclones may show that such fronts may not vary more than 10 or 20 degrees from this position without altering significantly the isohyetal pattern. This restricts the transposition of this type of storm which therefore may not be orientated more than 20 degrees. Such a storm is transposed by using a tracing of the isohyetal pattern and superimposing it over the project basin to give the maximum depth without varying the orientation more than the permissable amount of 20 degrees. Maximisation of the storm and allowances for moisture barriers or heights can proceed by the usual adjustment methods. As a general rule, storms should not be orientated beyond 20 degrees.

Superposition of a storm has advantages in that not only meteorological factors are considered, but topography, moisture inflow, and air mass trajectories are taken into account and a degree of professional judgement is used. For these reasons some consider the method to give sounder estimates of probable maximum precipitation than the simple transfer of the depth, duration, area values. It follows that estimates from the transposed isohyetal patterns will be lower than those obtained from depth duration area data. The latter virtually assumes that the maximum storm isohyets follow the outline of the river basin and is centred within it.

17.8 Fit factors. The project basin itself may possess an irregular shape or unusual orientation. It is inappropriate to transpose the maximum depth from a storm of the same area but different shape to the basin without modification. A fit factor is needed and is obtained by superimposing the basin outline over the isohyets of the critical storm. By planimetering the isohyets within the basin, a rainfall depth is obtained which can be compared with the highest average rainfall depth for the same area of the storm. The ratio of these is the fit factor which may now be used with the transposed and adjusted rainfalls for each duration. Due regard may be paid to the possible orientation of the storm in this procedure.

17.9 Transposition and maximisation of storms in mountainous areas. When the topographical influence on a storm is limited to the depletion of the inflowing moisture, its transposition to similar areas and its maximisation represent few problems. If, however the slope, altitude, orientation, etc. of a mountain range affects the flow pattern of the storm and causes intensification of precipitation and spillover effects, transposition and maximisation are not so simple. If high efficiency storms have occurred within the valley, the best solution is to maximise these '*in situ*' storms as discussed in paragraph 17.2. However, if there is an absence of data on intense storms in the valley other solutions have to be found and transposition may be considered. Also, it skilled staff are available, a storm model may be designed and the vertical lift, intensification, and spillover computed. Reports of the hydrometeorological sections of the U.S. Weather Bureau and of the Resources Agency of California illustrate successful studies of this type in the mountainous regions of the United States and give the necessary detail. [2,6]

There are ways of using and transposing actual storms which occurred outside the project basin. Firstly, storms within the same meteorologically homogeneous zones may be transposed between areas of similar topography. Storm orientation and fit factors must be considered. Expert judgement on the suitability of storms and the similarity of the basins for transposition is required. Secondly, the storm precipitation depths may be expressed as percentages of the mean precipitation for a month, season, or year. These isopercentals can be transposed to other basins using the principles of transposition already laid down. That is, by using moisture adjustments for barriers and maximum dew points as well as orientation or fit factors when necessary. In their new locations, the adjustment isopercentals are applied to the local mean isohyetal map for the month, season, or year, corresponding to the period chosen in computing isopercentals. The result is the maximum depths of precipitation for the transpos di storm in the project basin.

If frequency analyses of rainfall data have been made and, perhaps, expressed in terms of isohyetal maps of say, the once-in-2-yr 24-hr precipitation or the once-in-10-yr 72-hr precipitation, instead of using percentages of the mean precipitation, the storm depths may be expressed as a percentage or proportion of these parameters. These proportions may be transposed to a project catchment within the meteorologically homogeneous zone using the principles of transposition and maximisation and the deduced values applied to the same recurrence interval isohyetal patterns of the project basin. These isohyetal patterns are virtually used to express the topographic characteristics of the catchment.

The isopercental method can be simply illustrated by using examples of gauge rainfalls. For example, if 25 in. fell in 24 hr at a site where the average rainfall was 45 in. then this storm contributes 25/45 or 56% of the average annual rainfall. In another catchment area with maximisation and transposition factor of 1·2, the average annual rainfall is 40 in. and the transposed storm contributes $56 \times 1{\cdot}2$ or 67%. That is $40 \times 67\%$ or 26·8 in. This represents the probable maximum value for the storm in question. Similarly, using recurrence intervals, if the once-in-2-yr 24-hr rainfall were 2·00 in., the storm contributes 25/2·00 or 12·5 times this, and, on transposition and maximisation, should give $1{\cdot}2 \times 12{\cdot}5 = 15{\cdot}00$ times. On the project basin the once-in-2-yr 24-hr rainfall is 1·80 in. and $1{\cdot}80 \times 15{\cdot}0$ or 27·0 in. may be expected from the storm in the new location with maximum moisture inflow.

REFERENCES

[1] WIESNER C.J., (1965). Hydrometeorology and river flood estimation. Discussion, *Proc. Inst. Civ. Engrs.* **31**, 79–86.

[2] KNOX J.B., (1961). A review of the water vapor convergence method of estimating maximum possible orographic precipitation. *Dept. Water Resources, California.*

[3] U.S. Army Corps of Engineers. (To date) Storm Rainfall in the United States.

[4] World Meteorology Organisation (1969). Manual for depth, duration area analysis of storm precipitation. W.M.O. No. 237 TP 129.

[5] U.S. Weather Bureau (1949). Representative twelve-hour dew points in major U.S. storms east of the continental divide. *Hydromet. Report. No. 25A.*

[6] U.S. Weather Bureau (1961). Interim report, Probable Maximum Precipitation in California. *Hydromet. Report No.* 36.

CHAPTER EIGHTEEN

Estimating the P.M.P.—Generalized data—Empirical formulae—Statistical methods

18.1 Highest observed precipitation depths. In determining the P.M.P. in project catchments, it is necessary to consider, for various durations, the greatest storm rainfall depths which have been recorded, firstly within the catchment, and secondly, within the homogeneous meteorological zones. These are adjusted and/or transposed by methods already discussed. In some regions there is a short history of storms, or those adequately documented are not highly efficient. Therefore, it is useful to compare the available records with those of other areas, or continents. A list of the greatest recorded rainfall depths for different durations both at rain gauges, or point rainfalls, and from storm analyses over specified areas, depth, duration area, data, is a guide to P.M.P. estimations.

Precipitation gauge records are available throughout the world and the maximum recorded values for various durations are given in Table 18.1 and shown in Fig. 18.1. [1]

Table 18.1. *World's greatest observed rain gauge depths (as at 1965).*

Duration	*Depth* (in.)	*Location*	*Date*
1 min	1·23	Unionville, Md	July 4, 1956
8 min	4·96	Füssen, Bavaria	May 25, 1920
15 min	7·80	Plumb Point, Jamaica	May 12, 1916
20 min	8·10	Curtea-de-Arges, Roumania	July 7, 1889
42 min	12·00	Holt, Mo	June 22, 1947
2 hr 10 min	19·00	Rockport, W. Va	July 18, 1889
2 hr 45 min	22·00	D'Hanis, Tex (17 m. NNW)	May 31, 1935
4 hr 30 min	30·8	Smethport, Pa.	July 18, 1942
9 hr	42·79	Belouve, La Reunion	Feb 28, 1964
12 hr	52·76	Belouve, La Reunion	Feb 28–29, 1964
18 hr 30 min	66·49	Belouve, La Reunion	Feb 28–29, 1964
24 hr	73·62	Cilaos, La Reunion	Mar 15–16, 1952
2 days	98·42	Cilaos, La Reunion	Mar 15–17, 1952
3 days	127·56	Cilaos, La Reunion	Mar 15–18, 1952
4 days	137·95	Cilaos, La Reunion	Mar 14–18, 1952
5 days	151·73	Cilaos, La Reunion	Mar 13–18, 1952
6 days	159·65	Cilaos, La Reunion	Mar 13–19, 1952
7 days	161·81	Cilaos, La Reunion	Mar 12–19, 1952
8 days	162·59	Cilaos, La Reunion	Mar 11–19, 1952
15 days	188·88	Cherrapunji, India	Jun 24–Jul 8, 1931
31 days	366·14	Cherrapunji, India	July, 1861

Duration	*Depth* (in.)	*Location*	*Date*
2 months	502·63	Cherrapunji, India	June–July, 1861
3 months	644·44	Cherrapunji, India	May–July, 1861
4 months	737·70	Cherrapunji, India	April–July, 1861
5 months	803·62	Cherrapunji, India	April–Aug 1861
6 months	884·03	Cherrapunji, India	April–Sep, 1861
11 months	905·12	Cherrapunji, India	Jan–Nov, 1861
1 yr	1041·78	Cherrapunji, India	Aug 1860–July 1861
2 yr	1605·05	Cherrapunji, India	1860–1861

The occurrence of intense rainfalls are due to a combination of exceptional conditions such as the presence of a high dew point air mass together with an efficient storm mechanism, and topography favourably located to the inflowing moist air. All three factors are often involved. Topography, in particular, gives persisting and high rates of precipitation for durations of 24 hr and longer. For example, at Cilaos, elevation 3937 ft, La Reunion, the winds associated with tropical cyclones force moist unstable air upslope into the funnel of a steep valley which rises to 10 000 ft. Here the world's highest rainfall for durations of 1 to 8 days were recorded in 1952. However as there is a large orographical component they may not be applicable to other sites. Table 18.1a gives the next intense rainfalls which have been reported and may not be so limited.

Table 18.1a. *Second highest world's observed raingauge depths*

Duration	*Depth* (in.)	*Location*	*Date*
21 hr	41·7	Kadena Air Force Base, Okinawa	Sept 8, 1956
1 day	45·99	Baguio, Phillipines	Aug 14–15, 1911
24 hr	49·13	Paishik, Taiwan	Sept 10–11, 1963
48 hr	63·78	Paishik, Taiwan	Sept 10–11, 1963
2 days	82·11	Bowden Pen, Jamaica	Jan 22–23, 1960

Short duration intense rainfalls occur over small areas and are probably due to the sudden collapse of strong vertical currents which support considerable quantities of liquid water. These localised storms are often referred to as cloud bursts and are recorded only when they pass directly over a rain gauge. Even so, the catch of the raingauge is affected by local turbulence, variations in vertical velocities, and rapid, short term fluctuations in the moisture flowing into the area above the raingauge. It is doubtful whether these fluctuations can be accurately measured, let alone extrapolated to their probable maximum. In addition, most of the heavy point rainfall has occurred in high dew point air masses with little or no maximization necessary on this score. Also, in most cases, depletion of moisture with height is compensated by orographical intensification with no adjustment needed for elevation. For these reasons, point rainfalls are not usually maximised and the values of Table 18.1 and the envelope of Fig. 18.1 are accepted as estimations of maximum cloud burst rainfall. The envelopment of the world's greatest observed raingauge depths plotted on a log scale against the log of their duration is shown in Fig. 18.1 and is expressed by the equation $P = 16{\cdot}6D^{0{\cdot}475}$ (18.1) where D is the duration in hours and P the precipitation in in. over the duration D.[1]

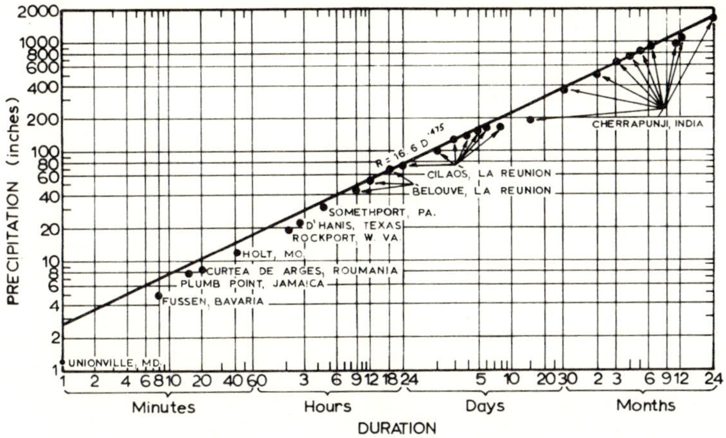

Fig. 18.1. World's greatest observed rain gauge depths.

The probable maximum point rainfall derived from this equation is given in Table 18.2 and is used for estimates of cloud burst rainfall over limited areas, with extreme dew point of 78°F and without adjustment for elevation. The data should be used only in areas with climates similar to the stations of Table 18.1. Rainfalls of comparable magnitude are unlikely in the colder climates and in elevated catchments protected by mountains where the air masses have low water contents. Even in equatorial climates outside the cyclone belt, the absence of efficient storm mechanisms may make their occurrence unlikely. In all cases, an exhaustive check of the raingauge and pluviograph data in the region is necessary to give a first base for estimates of extreme rainfall.

Table 18.2. *Probable maximum cloudburst rainfall.*

Duration (min)	*Depth* (in.)	*Duration* (hr)	*Depth* (in.)
5	4·76	1	16·6
10	6·8	2	23·4
15	8·3	3	28·7
20	9·6	4	33·2
30	11·8	5	37·0
45	14·4	6	40·6

Although the above estimates apply to raingauge depths, it is possible to extrapolate to wider areas by using cloudburst depth, area relationships. There is a more rapid decrease in the average rain depth with area than for other storms and Fig. 18.2, derived in the United States from depth, area data of cloudburst rainfall, can be used to extrapolate the data of Table 18.2.[2] From this the 1-hr cloudburst rainfall of 16·6 in extrapolated to 150 square miles is approximately 61% of the point value or 10·1 in.

No satisfactory temporal pattern has been observed for cloudburst rainfall and the chances of the heaviest portion occurring at the beginning, middle, or end of the storm duration are about equal. Thus the most critical distribution of cloudburst rainfall for the basin may be assumed.

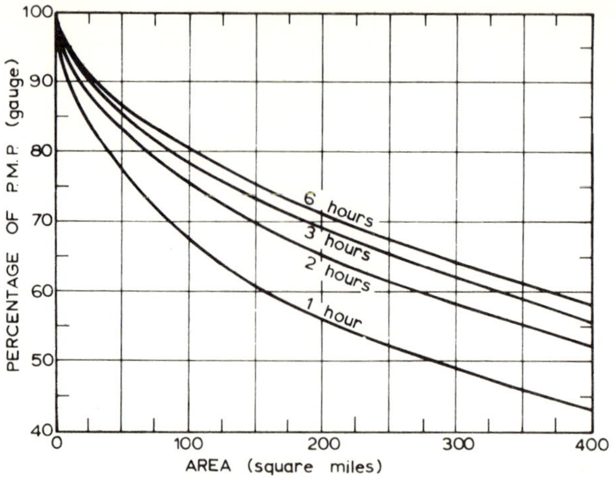

Fig. 18.2. Relation between depth and area for cloudburst rainfall in the U.S.

18.2 Highest values of depth, duration, area data. The areal and temporal pattern of storms and their maximum values are best obtained from depth, duration, area analyses, discussed in Chapter 11. However because of the more sophisticated techniques and the great amount of work involved in the storm analyses which yield depth, duration, area data, a list of the highest values of D.D.A. data for all countries, similar to Table 18.1, is not available. By far the greatest amount of this data comes from the United States where over 700 storms have been analysed in the required fashion and the maximum values are given in Table 18.3. However, interest in the method is spreading and limited information is available from Europe, Asia, Australia and Africa. None of the accessible D.D.A. data have shown values exceeding those of the United States and for this and other reasons, it may be assumed that the maximum depth, duration, area data for the United States, shown in Table 18.3, represents a useful upper limit for all countries. Although the values in Table 18.3 are believed to be the highest recorded for given durations and areas, future analyses from other parts of the world may well give higher values. The United States values have been revised several times since their first publication and, with the occurrence of more intense and efficient storms within areas now better served with recording stations, may again be revised in the future. This should be remembered when using earlier prepared diagrams and formulae based on the data available at that time. [4,5] Table 18.3 contains the latest data and should be used rather than derived figures and equations of earlier publications which do not take account of this revised data.

18.3 Maximisation and transposition of highest DDA data. Each of the storms contributing to Table 18.3 has a representative surface or 1000 mb dew point which is quoted. Although this indicates the moisture inflow into the storms which cover moderate areas, say up to 5000 square miles, it is doubtful whether it specifies the moisture supplied to larger areas. There are not only variations in dew point with distance, but the diverse topography, differing precipitation patterns, and the distinct orientation of the storms in relation to moisture inflow require special assessment of the moisture factors in the larger storms

Table 18.3. *Maximum Depth, Duration, Area data for the United States* (1960).

Area (*square miles*)	*Duration* (hr) 1	3	6	12	18	24	36	48	72
10	8·8	18·8b	24·7a	29·8b	36·3c	38·7c	41·8c	43·1c	45·2c
100	6·0b	14·7b	19·6b	26·3c	32·5c	35·2c	37·9c	38·9c	40·6c
200	5·5b	13·5b	17·9b	25·6c	31·4c	34·2c	36·7c	37·7c	39·2c
500	4·7b	11·8b	15·4b	24·6c	29·7c	32·7c	35·0c	36·0c	37·3c
1000	3·9b	10·0b	13·4b	22·6c	27·4c	30·2c	32·9c	33·7c	34·9c
2000			11·2b	17·7c	22·5c	24·8c	27·3c	28·4c	29·7c
5000			8·1bd	11·1b	14·1b	15·5c	18·7e	20·7e	24·4e
10000			5·7d	7·9f	10·1g	12·1g	15·1e	17·4e	21·3e
20000			4·0d	6·0f	7·9g	9·6g	11·6e	13·8e	17·6e
50000			2·5gh	4·2i	5·3g	6·3g	7·9g	8·9g	11·5j
100000			1·7h	2·5hk	3·5g	4·3g	5·6g	6·6j	8·9j

Storm	*Date*	*Location*	*Dew pt.*
a	July 17–18, 1942	Smethport, Pa.	74
b	Sept 8–10, 1921	Thrall, Tex.	77
c	Sept 3–7, 1950	Yankeetown, Fla.	76
d	June 27–July 4, 1936	Bebe, Tex.	78
e	June 27–July 1, 1899	Hearne, Tex.	75
f	April 12–16, 1927	Jefferson, Parish La.	72
g	March 13–15, 1929	Elba, Ala.	67
h	May 22–26, 1908	Chattanooga, Okla.	74
i	April 15–18, 1900	Eutaw, Ala.	66
j	July 5–10, 1916	Bonifay, Fla.	76
k	Nov. 19–22, 1934	Millry, Ala.	69

or even for all storms. However, by methods already discussed, the storms can be maximised and transposed within their meteorologically homogeneous zones using the dew points within the zones. Together with other data, the U.S. Hydrometeorological Section has thus constructed generalised maps to give the P.M.P. for various areas and durations. [6–8] A sample map is shown in Fig. 18.3.

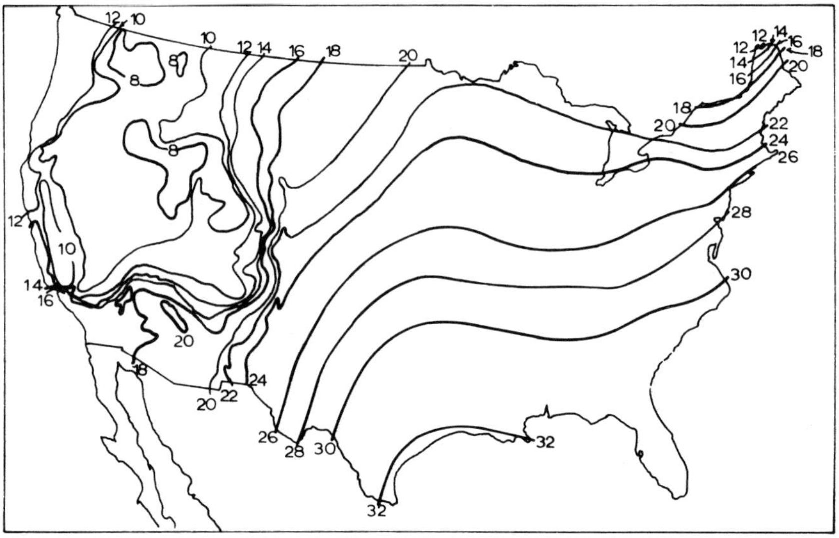

Fig. 18.3. P.M.P. for 6 hr and 10 square miles in the U.S.

With appropriate depth, duration, area relationships determined for the various zones from the storms which have occurred there, tables and graphs have been produced to deduce P.M.P. for other areas and durations. [6–8] The data applies to the regions for which it was determined and requires examination before it can be used in other areas. Adjustments for change in elevation are made when depletion of moisture with height is not balanced by orographical intensification, but are disregarded when one factor balances the other.

With a view to wider transposition, the small area and short duration storms have been more closely examined. Most of these are convergent types of models with inflow from high dew point air masses and strong vertical currents which are related to the convergent radial inflow at the bottom and divergent outflow at the top. Thunderstorms are typical examples of this type of mechanism and the intense D.D.A. values for short durations (less than 24 hr) and small areas (less than 1000 square miles) are often referred to as thunderstorm rainfall depths and the convergent mechanism as the thunderstorm model. This type of storm is not limited in its extent of transposition as are other systems like the tropical or frontal cyclones which have latitudinal limits. There are few areas in the world where thunderstorms or convergent models cannot occur and consequently such models are considered to be universally transposable. The United States D.D.A. data for durations less than 24 hr and areas less than 1000 square miles is accordingly used in other countries[9] either in its original form in Table 18.3, or maximised to a dew point of 78°F.

Table 18.4 shows maximum gauge rainfall together with depth, duration, area values raised to a dew point of 78°F. These values are approximately 10% higher than the original values of Table 18.3.

Table 18.4. *Probable maximum depth, duration, area data (dew pt 78°F)*

Area (*square miles*)	*Duration* (hr) 1	3	6	12	18	24
Gauge	16·6	28·0	38·9	54·0	65·5	75·1
10	9·3	19·8	30·0	31·3	40·0	42·6
100	6·3	15·5	20·6	29·0	35·8	38·8
200	5·8	14·2	18·8	28·2	34·6	37·7
500	4·9	12·4	16·2	27·1	32·7	36·0
1000	4·1	10·5	14·1	24·9	30·2	33·3
2000			11·8	19·5	24·8	27·3
5000			8·5	11·7	14·8	17·1

For short durations and small areas, the P.M.P. may be determined from Table 18.4 with an appropriate adjustment to the maximum 12-hr dew point of the project basin. If the 1000 mb dew point of the basin is used, there is no allowance for elevation. If project basins are elevated or there are moisture barriers in the path of the significant air masses, this effect should be assessed. If moisture depletion is not balanced by orographical intensification then allowances must be made for elevation. This is an additional adjustment to that of maximisation. Both factors are taken into account by using the maximum dew point at the height of the moisture barrier or at the catchment elevation to adjust the depths of Table 18.4. If intensification balances depletion, the adjustment factor is derived from the maximum 1000 mb dew point of the project catchment.

18.4 The use of United States data in other countries. In many countries there is an understandable reluctance to use the data from other continents. However, as DDA data is lacking outside North America it is useful to assess the extent of application of the U.S. D.D.A. data. For instance, the larger area storms in islands such as Great Britain and Tasmania show lower depths than in North America, although the small area, short duration storms are comparable. On the other hand, the continents of Europe and Asia are climatologically similar to North America and, perhaps, storm depths for all durations and areas might be expected to equal or even exceed the U.S. values. However, until storm analyses have given higher values, the United States D.D.A. values may be taken as representing extreme world rainfalls. The following reasons support this contention.

(*a*) North America extends over a wide range of latitudes from the Tropics to the Arctic. All known types of air masses occur in these regions and possess extreme properties which are comparable with those of other continents.

(*b*) All known types of storms occur over the North American land mass and adjacent waters.

(*c*) The shape of the North American continent and its topography of high contrast with prominent north to south ridges allow the exchange of polar and tropical air masses as well as the development and modification of contrasting continental and maritime air masses. The reported extreme conditions suggest that storms involving such air masses will be of high efficiency.

(*d*) The observed world's maximum point precipitation depths show that North American stations are represented and do not depart far from the highest recorded. For durations greater than 24 hr, special conditions apply.

Acceptance of the D.D.A. depths of Table 18.3 or the restricted maximised values in Table 18.4 as the world's probable maximum precipitation depths gives a simple quick method of preliminary estimation of P.M.P. in other regions. Valid allowances for both dew point and elevation in a specific area are made to give somewhat lower depths. Also, this procedure serves as an overall check on the other methods used or gives a comparison of storm efficiencies within the different regions. The large amount of work involved in a local study may not be justified in the early design stage. It is confidently expected, however, that such a study would give lower values than Tables 18.3 and 18.4. In countries outside the U.S. the small area, short duration storms, or the thunderstorm rainfalls, show the closest fit to Tables 18.3 and 18.4, and are accepted as estimates of the P.M.P. with appropriate dew points adjustments.

18.5 Generalised estimates of convergent or thunderstorm rainfalls. The universal occurrence and transposability of the convergent or thunderstorm model together with the fact that, in many countries, the U.S. D.D.A. values have been approached for short duration, small area storms, have led to the use of Tables 18.3 and 18.4 with the maximum recorded dew points at station level to give values of P.M.P. in these countries. From the adjusted depths, isohyets of P.M.P. are drawn for the respective durations and areas. These maps are essentially dependent on the thunderstorm rainfalls as well as the maximum 1000 mb dew points and the contours of the country concerned. In Australia in 1951, this method was used to give generalised estimates of P.M.P. for various areas and durations, one example being Fig. 18.4.[10] These maps may be used in conjunction with rainfall frequency analyses. For a design value a proportionate part of the P.M.P. is sometimes accepted.

A conservative practice in Australia restricts the use of the U.S. data to durations of 6 hr or less, for areas under 200 square miles, and along the coast where, local storms of the required specifications have approached the U.S. values. However, the possibility of similar storms of larger area and longer duration occurring further inland should not be disregarded and in the absence of other information, the U.S. D.D.A. data can be used. As pointed out in paragraph 18.2, due to the revision of the D.D.A. values over the last 15 yr, subsequent on the occurrence of more intense storms and improved analyses, some of the earlier published data, diagrams, formulae and generalised maps give slightly different results from Table 18.3.[4,5] It is thus preferable to use the current U.S. data with appropriate adjustments unless it is known that the other aids are based on the most recent information. In any case, a detailed local study is to be preferred.

18.6 Estimating the P.M.P. from empirical formulae. Paragraph 1ᴗ.1 describes the world's highest recorded gauge rainfalls and gives the equation of their envelope in terms of duration as $P = 16{\cdot}6D^{0{\cdot}475}$. Studies of heavy rainfalls in other areas show considerable variations in the constant of proportionality but the exponent of D approximates to $0{\cdot}5$.[1]

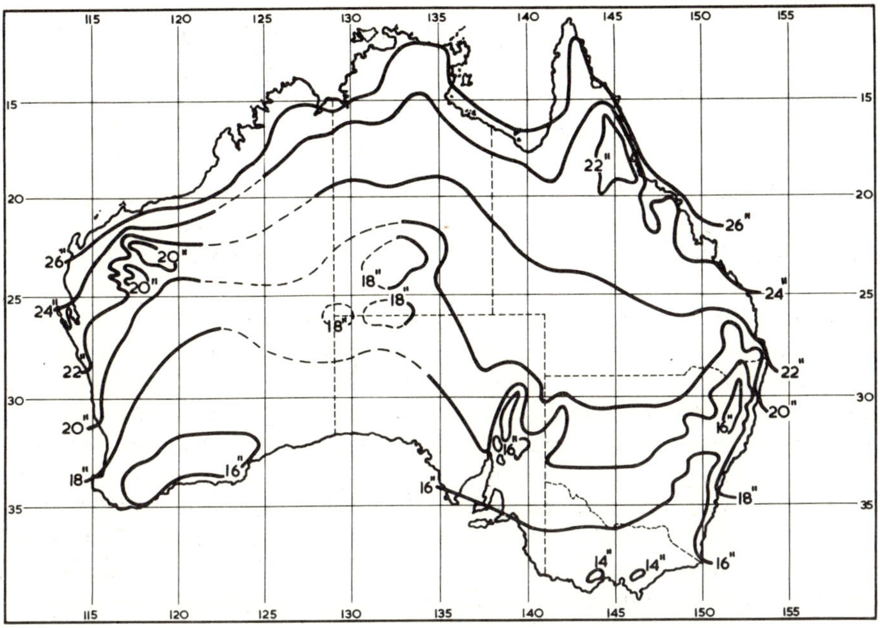

Fig. 18.4. Generalised estimates of P.M.P. for Australia.

Thus P varies as $\sqrt{D}$ for the world's highest gauge rainfalls, although for specific areas it is a good fit for only some of the durations. Long duration intense rainfalls require exceptional conditions including advantageous topography not achieved everywhere.

An examination of the extreme rainfall depths shows that they vary approximately inversely with the square root of the area over which the depth is determined. The data of Table 18.3 may be plotted as $P/\sqrt{D}$ against $\sqrt{A}$. The envelopment of this data is now seen as a rectangular hyperbola with the axes approximately parallel to and displaced from the original axes. An equation of the form

$$P = \sqrt{D}\,(a + b/\sqrt{A + c}) \qquad (18.2)$$

can be fitted to the highest recorded D.D.A. data. Using the available data at the time, Fletcher in 1950[5] estimated $a = 0{\cdot}5$, $b = 266$ and $c = 19{\cdot}2$. Changes in the D.D.A. data since that time has changed the value of these constants. An evaluation from Table 18.3 is $a = 0$, $b = 350$ and $c = 22$, while for intense storms in N.S.W., Australia, but for lower dew points than Table 18.3, $a = 1{\cdot}36$, $b = 85{\cdot}3$, and $c = 20$. [12]

Empirical formulae of this type represent local or world maximum values of precipitation and are useful for approximate estimations of P.M.P. Tests of the formula for various areas against the corresponding maximum observed data shows close envelopment for a restricted range of durations. [12] It is convenient, and perhaps a result of sampling, that the closest envelopment appears nearest that combination of durations and areas most useful in design work. More specific formulae can be derived within meteorologically homogeneous zones if desired but there is not much point using a formula if the original data is available and can be maximised and transposed to the specific catchment with all appropriate adjustments.

If the formula is derived from the highest recorded U.S. D.D.A. data in Table 18.3, or Table 18.4, it may be assumed that the formula applies to air masses with 1000 mb dew points approaching 78°F and persisting over a period of 12 hr. If the formula is used in a project basin it should be adjusted to the appropriate dew point with allowance for moisture barriers.

There are other formulae derived from storm models and have some theoretical basis but the constants of proportionality and integration are found from actual storms by empirical methods. These have been discussed under storm models in chapter 16.

18.7 Empirical relationships between the variables in particular valleys. It has been established that the rainfall intensity depends upon the inflow velocity, v, the moisture content, W, and a storm mechanism or convergence factor. Considering these factors for a given valley it may be concluded that well chosen observations of wind velocity and surface dew point represent meteorological conditions over the area more consistently than other observations, including rainfall. They also vary continuously and are more amenable to statistical treatment and extrapolation. Also, if intense storms have occurred and are related to the topography of the valley, it is likely that maximum values of the convergence factor have been reached in the valley's storm history. It is therefore possible to establish a rainfall formula directly relating rainfall intensity over the valley to wind velocity and the surface dew point and find the constants of the equation from the past intense storms. From estimates of the maximum values of dew point and wind velocities it follows that the maximum rainfall can be computed from the formula.

Platzmann did this for the Willamette basin in the United States [13] and found the best fit to the data an expression of the form

$$i = (a + bv^2)(c + dT_d). \tag{18.3}$$

Using 6-hr values for i, v, and T_d and five intense storms, he determined the constants by a least squares method to get $i = 17v^2 T_d \times 10^{-4}$ for the Willamette Valley. If it could be assumed that in this valley the maximum wind and dew points occurred concurrently in the selected storms and would occur in the probable maximum storm then the P.M.P. for 6 hr for the Willamette basin equals

$$17(v_{max})^2 T_{d\,max}\ 10^{-4}. \tag{18.4}$$

This approach is convenient in mountainous areas with complex topography and where there is limited data and manpower for elaborate storm model studies. Some of its assumptions are doubtful, for example the simultaneous occurrence of high dew points and wind velocities. The use of a quadratic term was justified by data and the combined effect of wind in orographical lift and frontal action. [13] Similar relationships can be established for other valleys using local data.

18.8 Estimating P.M.P. by statistical methods. Statistical methods are widely used to analyse precipitation gauge readings and, for design purposes, give the rainfall intensities for various durations and return periods. Some of the techniques are described in chapters 10 and 11. Raingauge data is taken as representing an area of 10 square miles and is converted to larger areas by using depth, area relationships. Due to the random nature of heavy storms or high precipitation rates, raingauge data are not successfully converted to areas beyond 400 or 500 square miles. For larger areas, a network of gauges is preferably used with the appropriate methods.

The fitting of rainfall data to extreme value distributions has caused controversy but gives consistency in the estimate of rainfall of selected frequencies and permits a more logical extrapolation of data. For most of the distributions used in Hydrology, the observed value X can be expressed in terms of its mean, $\bar{X}$, its standard deviation, σ, or coefficient of variation, $\sigma/\bar{X}$, and a frequency factor, K, defined from

$$X = \bar{X} + K\sigma \text{ or } X = \bar{X}\{1 + K(c \text{ of } v)\}. \tag{18.5}$$

The frequency factor, K, depends upon the statistical distribution used, the number of years of record, n, and the return period.[14] The relationship between these for the distribution employing the reduced variate or Gumbel distribution is shown in Fig. 10.12. This diagram and a knowledge of some of the factors of the distribution enables the others to be simply determined. The scales for $X - \bar{X}$, σ, and K can be multiplied by any convenient factor if necessary.

It is reasonable to expect that for the P.M.P. there is a value of K which will not be exceeded, say K_m. The P.M.P. may then be expressed as P.M.P. $= \bar{X} + \sigma K_m$.[15] The possibility that K_m may vary with other factors such as storm duration, location and mechanism has been considered and K_m has been empirically expressed in terms of storm duration and mean of the annual maximum rainfall as shown in Fig. 18.5[16] This relationship was set up by investigating the highest annual rain gauge readings in the United States for durations from 5 min to 24 hr. Values of K were determined for the highest recorded rainfalls for each duration as equal to

$$\frac{\text{Maximum observed rainfall} - \text{mean of annual series}}{\text{standard deviation}}$$

or

$$K_m = \frac{P_m - \bar{P}}{\sigma} = \frac{X_m - \bar{X}}{\sigma} \quad \text{— for each duration} \tag{18.6}$$

where m indicates highest value of the series.

The values of K_m obtained for each duration showed no geographical pattern but, when plotted against the mean of the annual maximum, show a decrease in K with increase in $\bar{P}$. By enveloping all the plotted values of K_m for each duration and embracing all available records, a diagram similar to Fig. 18.5 was obtained. This envelopment is a similar process to transposition. Use of the world's greatest observed values enable the upper position of the curves to be set. Tests of internal consistency also better define the curves of various duration and enable them to be interpolated, e.g. for 6 hr.

Figure 18.5 may be used to obtain an estimate of the P.M.P. at a point for various durations from the mean of the annual maximum rainfalls and their standard deviations.

$$\text{i.e. P.M.P.} = \bar{P} + K_m\sigma \tag{18.7}$$

It might be noted that stations with the same mean of the annual maximum precipitation may give different P.M.P.'s by virtue of differences in their standard deviations.

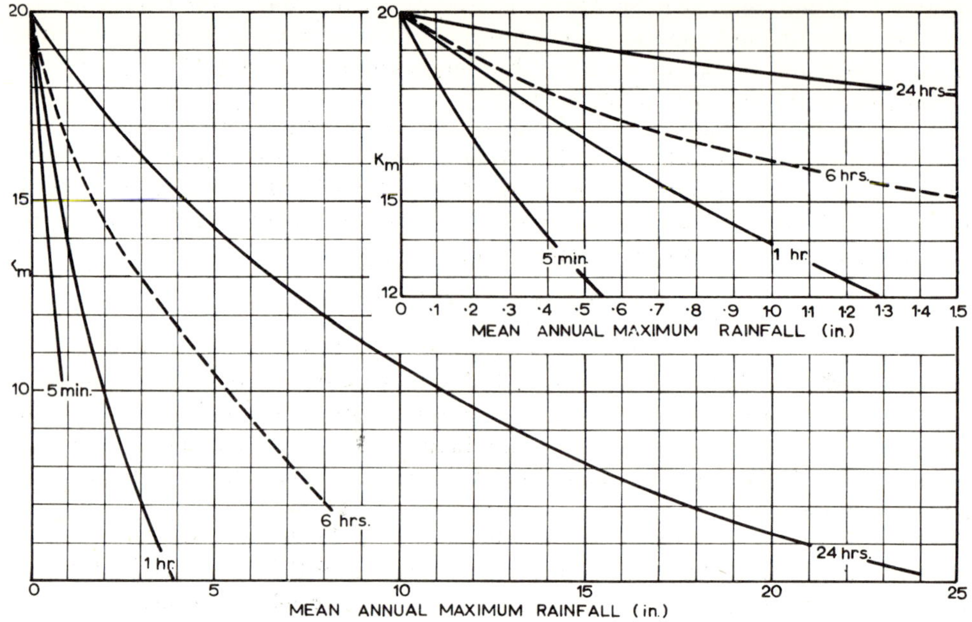

Fig. 18.5. Relationship between mean annual maximum rainfall, frequency factor K_m, and storm duration

Early work on this method indicated that the P.M.P. = $\bar{P} + 15\sigma$ was a reasonable approximation.[15] This was applied in some of the U.S. Hydrometeorological studies in conjunction with other methods.

The method has the advantages of taking account of the actual data, expressing it in terms of statistical parameters, and being easy to use. The disadvantages are that the local influences on the storm or its peculiarities are not assessed and these may be significant. However, the universal transposibility of the thunderstorm rainfalls and the use of the envelopment of the K_m values are similar processes and both might well be accepted for short duration, small area storms. In common with all statistical methods, its success depends upon the length and nature of the records used. The random occurrence of extreme falls may affect the evaluation of $\bar{P}$ and σ and so the estimation of P.M.P.

The ease of calculation of this method and its reliance on actual data suggests that it should be used in conjunction with the other methods.

REFERENCES

[1] PAULHAUS J.L.E., (1965). Indian Ocean and Taiwan rainfalls set new records. *Monthly Weather Rev.* 93, 331–335.

[2] U.S. Weather Bureau (1961). Generalised estimates of P.M.P. and rainfall frequency data for Puerto Rico and Virgin Islands. *Tech. Paper No.* 42.

[3] U.S. Corps of Engineers (To date). Storm Rainfall in the United States. First issued 1945.

[4] BERRY E.A., BOLLAY E. and BEERS N.R., (1945). *Handbook of Meteorology*, McGraw Hill, New York.

[5] FLETCHER R.D., (1950). A relation between maximum observed point and areal rainfall values. *Trans. Am. Geophys. Uni.* 31, 344–348.

[6] U.S. Weather Bureau (1947). Generalised estimates maximum possible precipitation over the United States east of the 105th meridian, areas of 10,200, and 500 square miles. *Hydromet. Report No.* 23.

[7] U.S. Weather Bureau (1956). Seasonal variation of the P.M.P. East of the 105th meridian for areas from 10–1000 square miles and durations of 6, 12, 24 and 48 hr. *Hydromet. Report No.* 33.

[8] U.S. Weather Bureau (1960). Generalised estimates of probable maximum precipitation for the United States west of the 105th meridian for areas to 400 square miles and durations to 24 hr. *Technical Note No.* 38.

[9] Bureau of Meteorology, Australia (1958). Conference on estimation of Extreme Precipitation, Melbourne.

[10] WALPOLE J., (1951). Maximum possible rainfall over Australia republished in Ref. 9 above.

[11] REICH B.M., (1963). Short duration rainfall intensity estimates and other design aids for regions of sparse data. *J. Hydrology* 1, 3–28.

[12] WIESNER C.J., (1967). Unpublished data. University of New South Wales.

[13] PLATZMANN G.W., (1948). Computations of maximum rainfall in the Willamette Basin. *Trans. Am. Geophys. Uni.* 29, 467–472.

[14] CHOW V.T., (1951). A general formula for hydrologic frequency analysis. *Trans. Am. Geophys. Uni.* 32, 231–237.

[15] HERSHFIELD D.M., (1961). Estimating the probable maximum precipitation. *Proc. Am. Soc. Civ. Engr. J. Hydraulics. Div.* 87, 99–106.

[16] HERSHFIELD D.M., (1965). Method for estimating probable maximum precipitation. *J. Am. W.W. Assn.* 57, 965–972.

Appendix 1

Recommended units for hydrometeorological elements. Commonly used alternative units and corresponding factors for conversion to recommended units are also shown.

Element (1)	*Recommended unit* (2)	*Alternative units* (3)	*Factor for conversion from alternative unit (3) to recommended unit (2)* (4)
Water-level (stage)	cm	ft	30·5
Stream discharge	m^3/sec	cfs	0·0283
Unit discharge	$m^3/sec/km^2$	cfs/mi^2	0·0103
Volume (storage)	m^3	ft^3	0·0283
		ac-ft	1230
		cfs-days	2450
Runoff depth	mm	in.	25·4
Precipitation	mm	in.	25·4
Precipitation intensity	mm/hr	in./hr	25·4
Snow depth	cm	in.	2·54
Snow cover, area	%		
Water equivalent of snowpack	mm	in.	25·4
Ice thickness	cm	in.	2·54
Evaporation	mm	in.	25·4
Evapotranspiration	mm	in.	25·4
Soil moisture	%, volume	%, weight (conversion depends on density)	
Soil-moisture deficiency	mm	in.	25·4
Sediment discharge	MT/day	tons/day	0·907
Sediment concentration	kg/m^3	ppm (conversion depends on density)	
Chemical quality	ppm		
Energy (heat)	cal (gramme)	Btu	252
Radiation	cal/cm^2	ly	
Radiation intensity	cal/cm^2/min	ly/min	
Sunshine	% possible	hrs (conversion depends on possible sunshine)	
Temperature	°C	°F	5/9 (°F − 32)
Wind speed	knots	mi/hr	1·95
	m/sec		0·868
Relative humidity	%		
Vapour pressure	mb	mm Hg	1·333
		in. Hg	33·86
Atmospheric pressure	mb	mm Hg	1·333
		in. Hg	33·86
Area	km^2	mi^2	2·59
		ac	0·00405
		ha	0·01

NOTE: Abbreviations used in the table are as follows:

ac – acre
Btu – British thermal unit
°C – degrees Celsius
cfs – cubic feet per second
cm – centimetre
°F – degrees Fahrenheit
ft – foot
ha – hectare
Hg – mercury
hr – hour
in. – inch
kg – kilogram
km – kilometre
ly – langley
m – metre
mb – millibar
mi. – mile
min – minute
mm – millimetre
MT – metric ton
ppm – parts per million by weight
sec – second

From W.M.O. Guide to Hydrometeorological Practices.

Appendix 2

Some letters and symbols used

a	air (e.g. T_a air temperature
C	cloud amount in fraction of sky covered
C_p and C_v	specific heats at constant pressure and volume
d	dew point (e.g. e_d s.v.p at dew point)
D	duration
e	base Napierian logarithms 2·71828
e	saturation vapour pressure s.v.p.
e_a	s.v.p. at T_a
e_d	s.v.p. at T_d
e_s	s.v.p. at T_s
e_w	s.v.p. at T_w
E	evaporation
i	precipitation intensity
K	transfer coefficient
K_H	heat transfer coefficient
K_W	moisture transfer coefficient
K_M	momentum transfer coefficient
L	latent heat of condensation or evaporation
m	rank in a series
m	mass
m_a	mass of dry air
m_w	mass of water vapour
M	snow melt
n	number, often of actual hours of sunshine
N	total, often total possible hours of sunshine
p	pressure
p_a	pressure of air
P	precipitation
P	probability
q	specific humidity
Q	runoff or volume of discharge
r	mixing ratio
r	reflectivity, albedo
R	gas constant or radiation
R_a	radiation at outer limit of atmosphere
R_b	radiation, back or terrestrial
R_c	radiation, inward or solar
R_n	radiation, nett
s	surface
S	storage
t	time
T	temperature
T_a	temperature, air
T_d	temperature, dew point
T_s	temperature, surface
T_v	temperature, virtual

T_w	temperature, wet bulb
T_r	return period
u, v, w	components of wind velocity in x, y, z directions
v	virtual
V	volume
W	precipitable water or liquid equivalent
W_e	effective precipitable water
x, y, z	components of displacement
$\bar{x}$	bar indicates average
β	Bowen ratio
Δ	slope of $e:T$ curve
δ	an increment of
γ	psychrometric constant
μ	10^{-4}cm
ϕ	latitude
π	3·1416
ρ	density
σ	standard deviation
Σ	summation
τ	shear stress
ω	angular velocity

Index